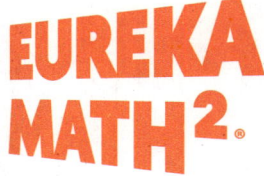

A Story of Units®

Units of Ten ▶ 1

APPLY

1 Counting, Comparison, and Addition

2 Addition and Subtraction Relationships

3 Properties of Operations to Make Easier Problems

4 Comparison and Composition of Length Measurements

5 Place Value Concepts to Compare, Add, and Subtract

Module **6** Attributes of Shapes · Advancing Place Value, Addition, and Subtraction

Great Minds® is the creator of *Eureka Math*®,
Wit & Wisdom®, *Alexandria Plan*™, and *PhD Science*®.

Published by Great Minds PBC.
greatminds.org

© 2021 Great Minds PBC. All rights reserved. No part of this work may be reproduced or used in any form or by any means—graphic, electronic, or mechanical, including photocopying or information storage and retrieval systems—without written permission from the copyright holder.

Printed in the USA
B-Print

3 4 5 6 7 8 9 10 11 12   QDG   27 26 25 24 23

ISBN 978-1-64497-641-8

EUREKA MATH²

Contents

Part 1: Attributes of Shapes

Topic A 3
Attributes of Shapes

Lesson 1 5
Name two-dimensional shapes based on the number of sides.

Lesson 2 9
Sort and name two-dimensional shapes based on attributes.

Lesson 3 13
Draw two-dimensional shapes and identify defining attributes.

Lesson 4
This lesson appears only in the *Teach* book ......

Lesson 5 17
Reason about the functionality of three-dimensional shapes based on their attributes.

Topic B 23
Composition of Shapes

Lesson 6
This lesson appears only in the *Teach* book ......

Lesson 7 25
Create new composite shapes by adding a shape.

Lesson 8 31
Combine identical composite shapes.

Lesson 9 35
Relate the size of a shape to how many are needed to compose a new shape.

Topic C 41
Halves and Fourths

Lesson 10 43
Reason about equal and not equal shares.

Lesson 11 47
Name equal shares as halves or fourths.

Lesson 12 53
Partition shapes into halves, fourths, and quarters.

Lesson 13 57
Relate the number of equal shares to the size of the shares.

Lesson 14 63
Tell time to the half hour with the term *half past*.

Lesson 15 67
Reason about the location of the hour hand to tell time.

Copyright © Great Minds PBC

Part 2: Advancing Place Value, Addition, and Subtraction

Topic D 73
Count and Represent Numbers Beyond 100

Lesson 16
This lesson appears only in the *Teach* book......

Lesson 17 75
Read, write, and represent numbers greater than 100.

Lesson 18 79
Count up and down across 100.

Lesson 19 83
Write totals for collections larger than 100 shown in various groups of tens and ones.

Topic E 89
Deepening Problem Solving

Lesson 20
This lesson appears only in the *Teach* book......

Lesson 21 91
Represent and solve *add to* and *take from* word problems.

Lesson 22 97
Represent and solve *add to* and *take from with start unknown* word problems.

Lesson 23 101
Represent and solve comparison word problems.

Lesson 24 107
Reason with nonstandard measurement units.

Lesson 25 111
Solve nonroutine problems.

Topic F 115
Extending Addition to 100

Lesson 26 117
Make a total in more than one way.

Lesson 27 121
Add two-digit numbers in various ways, part 1.

Lesson 28 129
Add two-digit numbers in various ways, part 2.

Lesson 29 133
Add tens to make 100.

Lesson 30 139
Make the next ten and add tens to make 100.

Lesson 31 143
Add to make 100.

Acknowledgments 147

Module 6
Topic A

FAMILY MATH
Attributes of Shapes

Dear Family,

Your student is studying geometry by exploring the defining attributes, or characteristics, of two-dimensional shapes.

Key Terms
parallel
rhombus
sphere
square corner
trapezoid

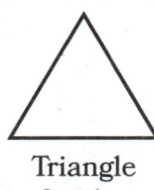
Triangle
3 sides

Quadrilateral
4 sides

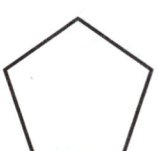
Pentagon
5 sides

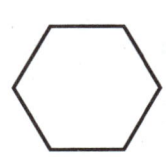
Hexagon
6 sides

Using the attributes, they name the shapes. Your student also draws, describes, and categorizes the two-dimensional shapes. For example, squares, rectangles, rhombuses, and trapezoids are quadrilaterals because they have 4 sides. They notice some shapes have square corners or parallel sides. Parallel sides do not touch, even if they were extended.

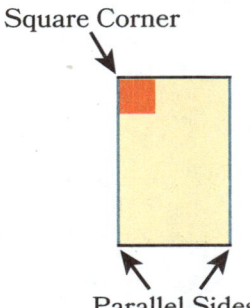

A rhombus is a closed shape with 2 pairs of parallel sides and 4 straight sides that are the same length.

A trapezoid is a closed shape with at least 1 pair of parallel sides and 4 straight sides.

Your student describes the attributes of three-dimensional shapes by identifying the face shapes, counting the number of faces, and describing their characteristics. For example, shapes with curved surfaces like cylinders and spheres can roll.

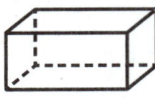
Rectangular prism

Cone

Triangular prism

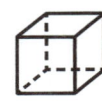
Cube

Cylinder

Pyramid

Sphere

At-Home Activities

Shapes Around Us

Encourage your student to observe the attributes of two-dimensional shapes in everyday life. For example, point out that the sides of a door are parallel, and two edges of the door meet to form a square corner. Have your student look for attributes in other shapes they see. For example, a sign for a school zone may be in the shape of a pentagon or a rhombus.

Stack or Roll?

Invite your student to find different objects that can be stacked, such as boxes and cans, and objects that can roll, such as cans and balls. For example, quarters can be stacked when flat or roll along their sides and cereal boxes can be stacked on all their sides. Then ask your student the following questions to discuss the attributes of the objects.

- "Which of the objects can both roll and stack? Why?"
- "What shapes are the objects?"
- "How might you group the objects?"

EUREKA MATH²   1 ▸ M6 ▸ TA ▸ Lesson 1

Name

1. Circle the shapes with **3** sides.

 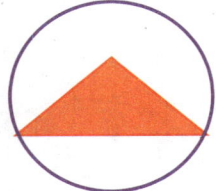  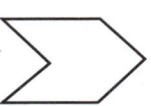  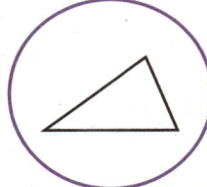

Draw a shape with **6** sides.
Sample:

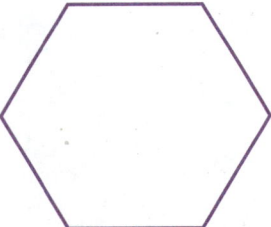

I count the number of sides on each shape.

I know shapes with 3 sides are called triangles.

I draw a shape with 6 sides.

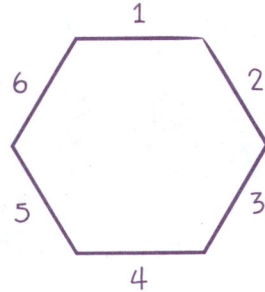

All sides have straight lines.

All sides connect, so my shape is closed.

REMEMBER

2. Count on to add.

$6 + 3 = 9$

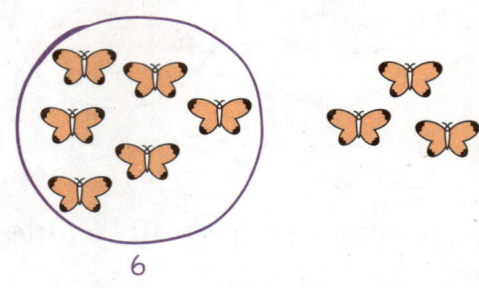

I see two groups of butterflies.

I circle a part I know.

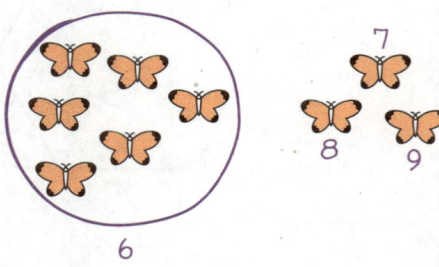

6

Then, I count on from 6.

6

3. Write the length of the pencil.

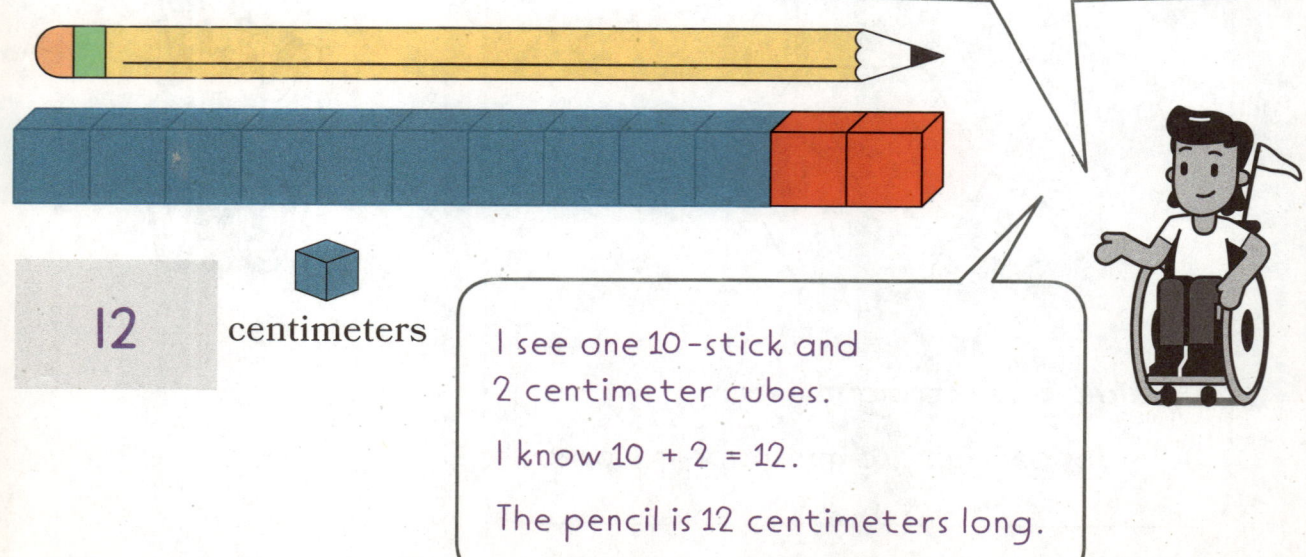

12 centimeters

I see one 10-stick and 2 centimeter cubes.

I know $10 + 2 = 12$.

The pencil is 12 centimeters long.

EUREKA MATH²

1 ▸ M6 ▸ TA ▸ Lesson 1

Name

1. Circle the shapes with **5** sides.

   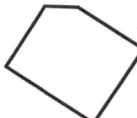  

2. Draw a shape with **3** sides.

3. Draw a shape with **4** sides.

7

REMEMBER

4. Count on from a part to add.

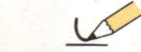

5. Write the length of the frog.

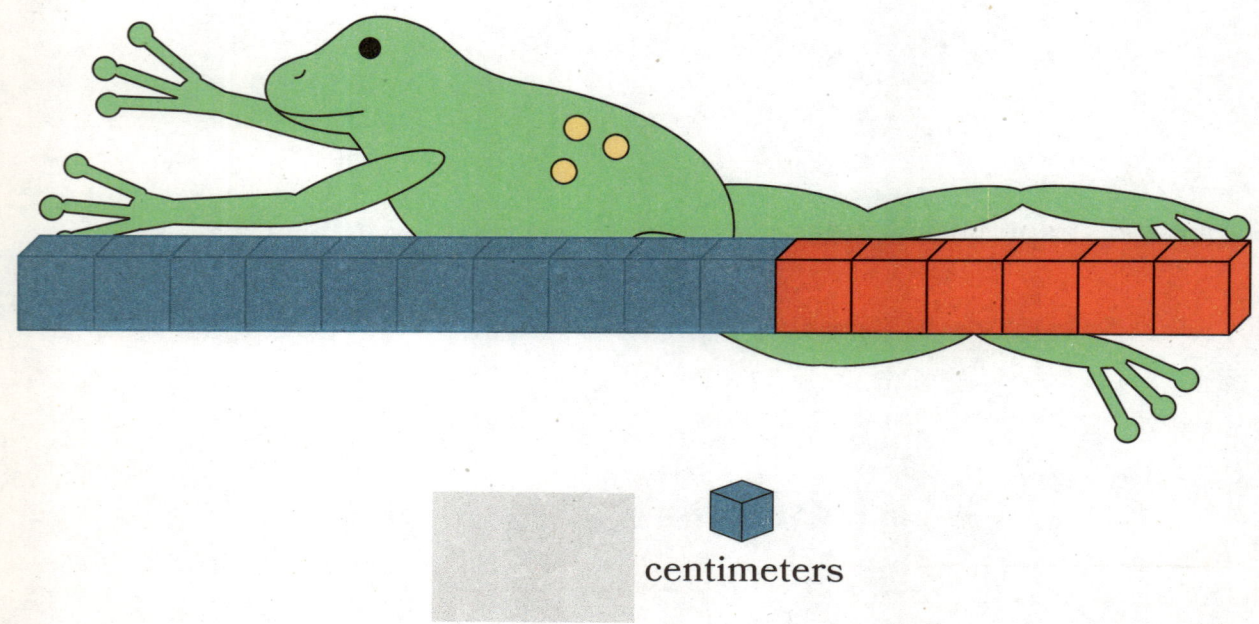

centimeters

EUREKA MATH²   1 ▸ M6 ▸ TA ▸ Lesson 2

2

Name

1. Count.

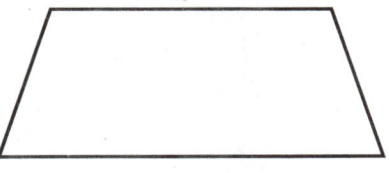

 sides

Does it have square corners?

Yes   or   (No)

 corners

Does it have parallel sides?

(Yes)   or   No

I know a square block will fit perfectly in a **square corner**.

I see the square fits in the corner of this rectangle.

I see the square does not fit in the corners of this trapezoid.

A **trapezoid** is a closed shape with 4 sides. I trace the sides as I count.

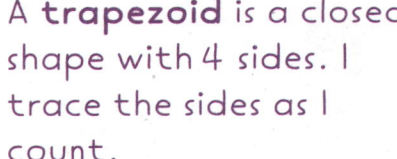

I know that **parallel** sides are across from each other. Parallel sides never touch.

I can use craft sticks or other straight objects to check for parallel sides.

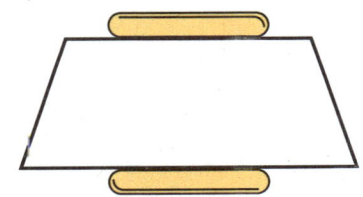

Copyright © Great Minds PBC

9

2. Circle the shapes with parallel sides.

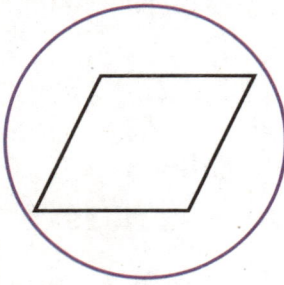 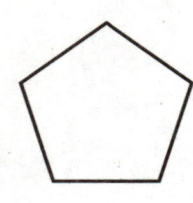 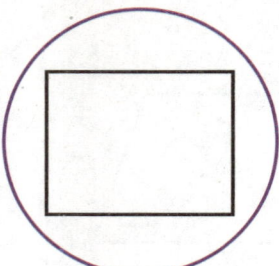

A **rhombus** has 4 sides that are all the same length. It has 2 pairs of **parallel** sides.

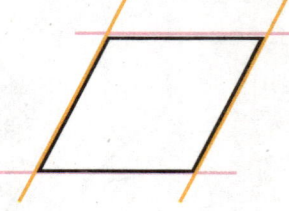

rhombus

A rectangle also has 2 pairs of **parallel** sides.

EUREKA MATH²   1 ▸ M6 ▸ TA ▸ Lesson 2

Name

1. Count.

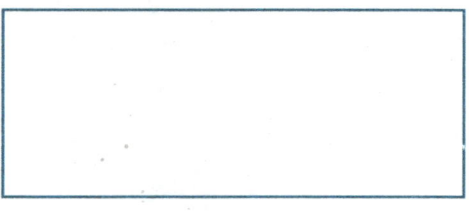

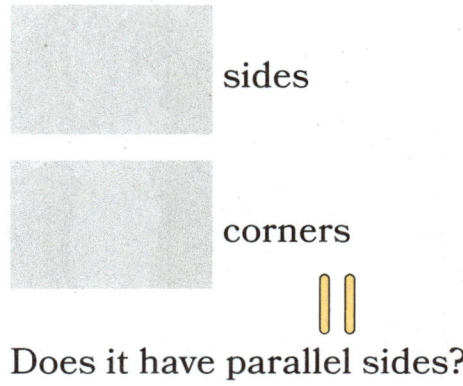

 Does it have square corners?

Yes   or   No

Does it have parallel sides?

Yes   or   No

Count.

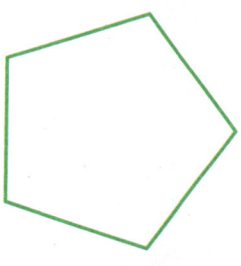

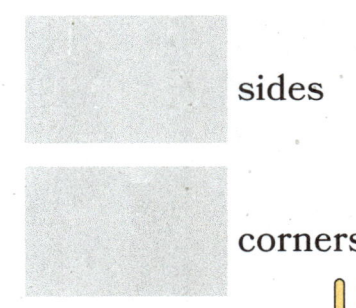

 Does it have square corners?

Yes   or   No

Does it have parallel sides?

Yes   or   No

2. Circle the shapes with parallel sides.

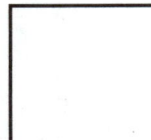   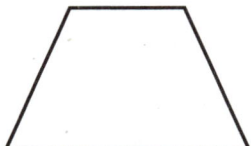   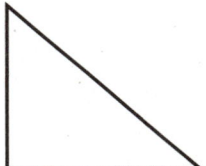

Name

1. Make a shape.

   Use the dots to draw straight lines.

   Rectangle

   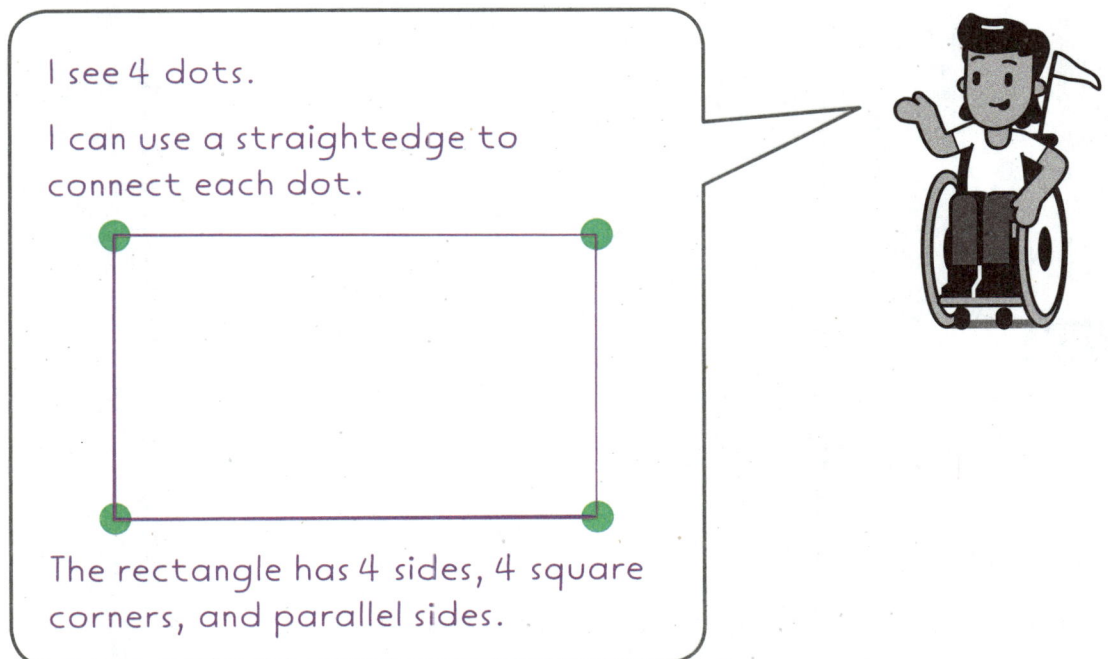

1 ▸ M6 ▸ TA ▸ Lesson 3　　　　　　　　　　　　　　　　EUREKA MATH²

REMEMBER

2. **Read**

   Baz finds 16 shells at the beach.

   He puts some back in the water.

   Now he has 11 shells.

   How many shells did Baz put back in the water?

   Draw　Sample:

   I read the problem. I read again. As I reread, I think about what I can draw.

   I draw 16 dots to show how many shells Baz finds.

   I circle the 11 he has now.

   I count how many shells he puts back in the water.

   Write

   $$16 - 11 = 5$$

   Baz puts   5   shells back in the water.

EUREKA MATH²   1 ▸ M6 ▸ TA ▸ Lesson 3

Name

1. Make shapes.

   Use the dots to draw straight lines.

   Triangle

   Hexagon

   Square

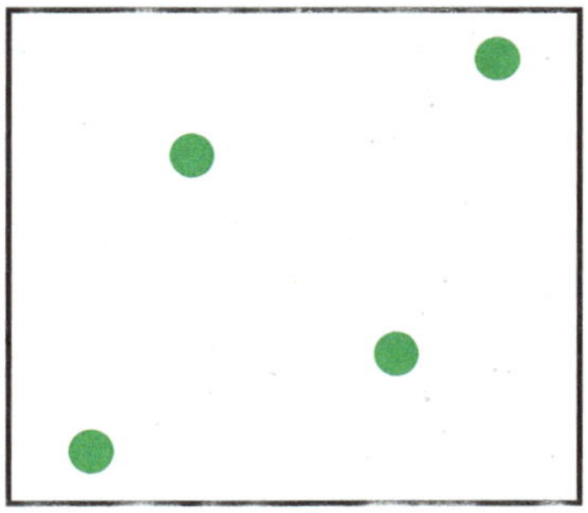

   Rhombus

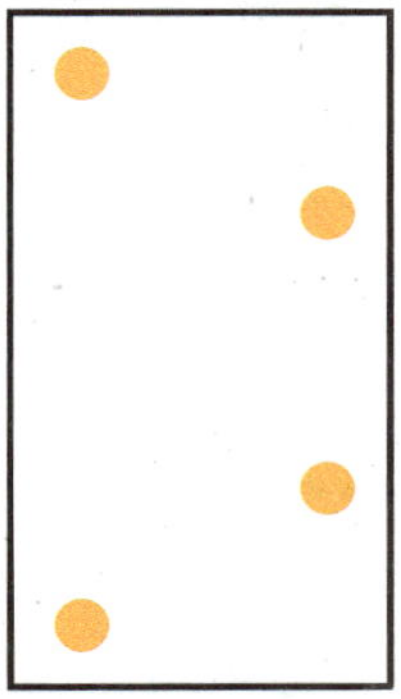

   Trapezoid

   Rectangle

REMEMBER

2. **Read**

   Zan has 19 pens.

   She gives some to her friends.

   Now she has 12 pens.

   How many pens did Zan give to her friends?

   Draw

   Write

   Zan gives _____ pens to her friends.

Name

1. Circle the shapes that roll.

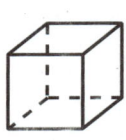 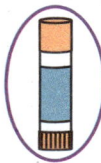  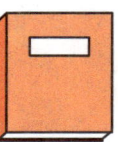   

2. Circle the shapes that stack.

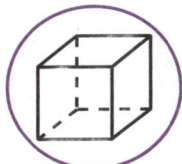 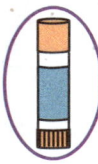     

I know that shapes with curves will roll. These shapes are curved.

I know that a shape with a flat face will stack. These shapes have a flat face.

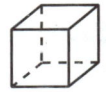 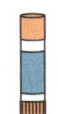  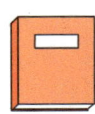  

3. Circle the shape that does **not** roll.

 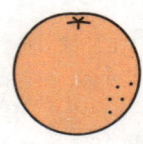  

4. Circle the shape that rolls but does **not** stack.

 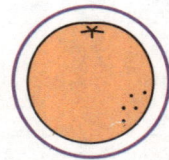  

> A shape without any curves does **not** roll.
> This shape does not have curves.
> It has flat sides and pointy corners.
>
>

> A shape with only curves will roll but will **not** stack.
> The orange is a sphere with only curves.
>
>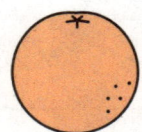

REMEMBER

5. Look at the picture.

   Color to show how many.

   Write the totals.

Animals We See   **Totals**

| 🐦 | 1 | 2 | 3 | 4 | 5 | 6 | 7 | 8 | 9 | 10 | | 3 |

| 🐟 | 1 | 2 | 3 | 4 | 5 | 6 | 7 | 8 | 9 | 10 | | 5 |

| 🦀 | 1 | 2 | 3 | 4 | 5 | 6 | 7 | 8 | 9 | 10 | | 4 |

I count each group of animals in the picture.

I touch and count the birds: 1, 2, 3.

The last number I counted is 3. There are 3 birds.

I can cross out each fish as I count. There are 5 fish.

I see there are 4 crabs.

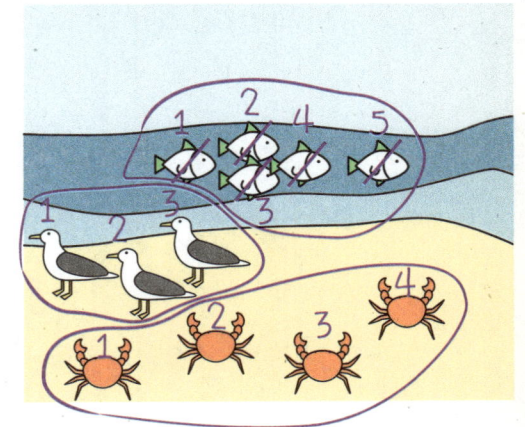

EUREKA MATH²   1 ▸ M6 ▸ TA ▸ Lesson 5

Name

1. Circle the shapes that roll.

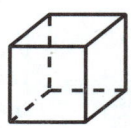   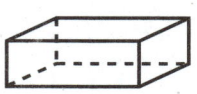  

2. Circle the shapes that stack.

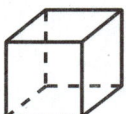   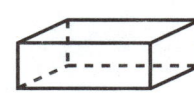  

3. Circle the shapes that do **not** roll.

4. Circle the shape that rolls but does **not** stack.

REMEMBER

5. Look at the picture.

   Color to show how many.

   Write the totals.

Animals We See   **Totals**

🐟	1	2	3	4	5	6	7	8	9	10

🐦	1	2	3	4	5	6	7	8	9	10

🦀	1	2	3	4	5	6	7	8	9	10

FAMILY MATH
Composition of Shapes

Module 6
Topic B

Dear Family,

Your student is exploring how smaller shapes can be combined to create new, larger shapes. A composed shape is made from putting 2 or more shapes together. Your student makes composed shapes by using combinations of shapes and drawings. They add to composed shapes and recognize the patterns that result.

Key Term
composed shape

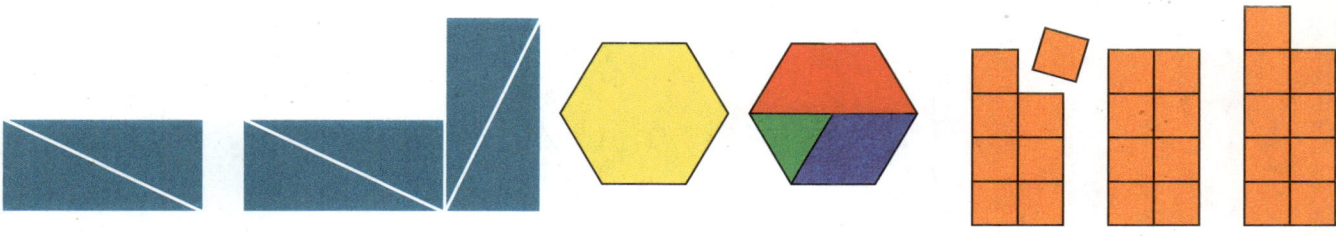

We composed a rectangle out of 2 triangles. We combined our rectangles to make a new composed shape.

This hexagon can be composed of a trapezoid, a triangle, and a rhombus.

Each time we add a shape, there is a pattern. When we add a square to this 6-sided figure, it becomes a 4-sided figure. When we add another square, it becomes a 6-sided figure again.

At-Home Activities

Search for Composed Shapes

Encourage your student to look for composed shapes around the home. For example, look at tiles in a kitchen or at a window with panes to explore how smaller shapes can be combined to make composed shapes. Ask your student questions like these to help them explore how smaller shapes can be combined to make composed shapes.

- "What shapes do you see?"
- "Can 2 small shapes be combined to make a composed shape? How about 4 small shapes?"
- "How many sides does the composed shape have?"
- "How many composed shapes do you see?"

Make Composed Shapes

Use masking tape or chalk to explore how a shape can be composed of other shapes. Start with a rectangular tabletop or section of sidewalk. Invite your student to use the tape or chalk to divide the surface into 2 smaller shapes. Ask your student questions like these to help them name and explore additional shapes.

- "Can you name the shapes?"
- "Could you divide the composed shape into even more shapes?"

Consider repeating the activity on a new surface or by creating different shapes on the same surface.

Copyright © Great Minds PBC

EUREKA MATH²   1 ▸ M6 ▸ TB ▸ Lesson 7

Name

1. Color a square. 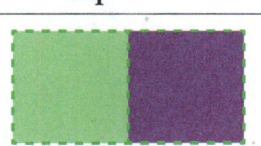 | Color a square.  | Color a square.

What is the last new shape? __square__   Sides **4**

I color the first square.
There are 2 squares.
It composed a rectangle.

I color the next square.
There are 3 squares.
Now it is a hexagon.

I color the next square.
There are 4 squares.
Now it is a square.

2. Color 4 triangles.

What is the new shape? __rhombus__   Sides 4

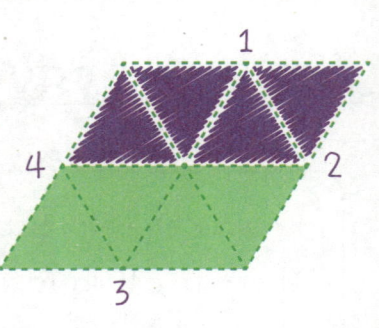

I color 4 triangles.
Now there are 8 triangles.
They compose a new shape with 4 sides.
It is a rhombus.

REMEMBER

3. Write as tens and ones.

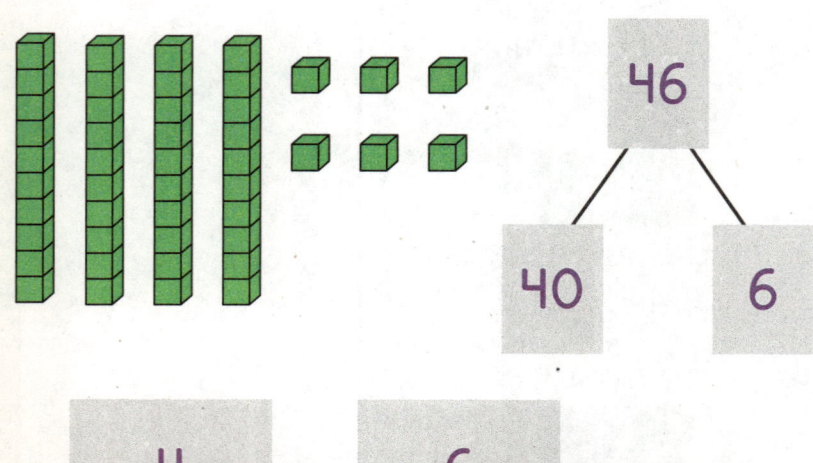

4 tens 6 ones

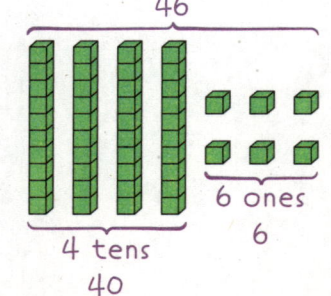

I see 4 tens 6 ones.
4 tens is 40. 6 ones is 6.
40 and 6 is 46.

4. Write each total.

Animals We See Totals

 | 1 | 2 | 3 | 4 | 5 | 6 | 7 | 8 | 9 | 10 | 11 | 12 | 13 | 14 | 15 | 10

 | 1 | 2 | 3 | 4 | 5 | 6 | 7 | 8 | 9 | 10 | 11 | 12 | 13 | 14 | 15 | 13

Which has more?

Circle.

Write the totals.

13 > 10

greater than

The number path for the frog is shaded to 10, so the total is 10.

The number path for the turtle is shaded to 13, so the total is 13.

The number path for the turtle is longer. There are more turtles than frogs because 13 is greater than 10.

Name _____

1. Color a triangle. | Color a triangle. | Color a triangle.

What is the last new shape? _____ Sides ____

2. Color 3 rhombuses.

What is the new shape? _____ Sides ____

REMEMBER

3. Write the total as tens and ones.

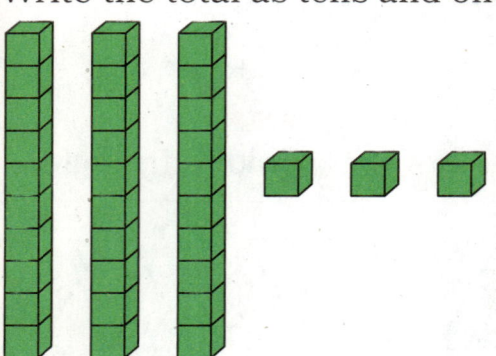

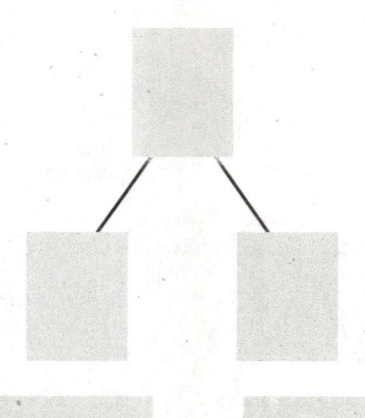

tens   ones

4. Write each total.

Animals We See   **Totals**

 | 1 | 2 | 3 | 4 | 5 | 6 | 7 | 8 | 9 | 10 | 11 | 12 | 13 | 14 | 15 |

 | 1 | 2 | 3 | 4 | 5 | 6 | 7 | 8 | 9 | 10 | 11 | 12 | 13 | 14 | 15 |

Which one has more?

Circle.

Write the totals.

___ > ___

greater than

Name _____

1. Color 1 triangle. | Color 2 triangles. 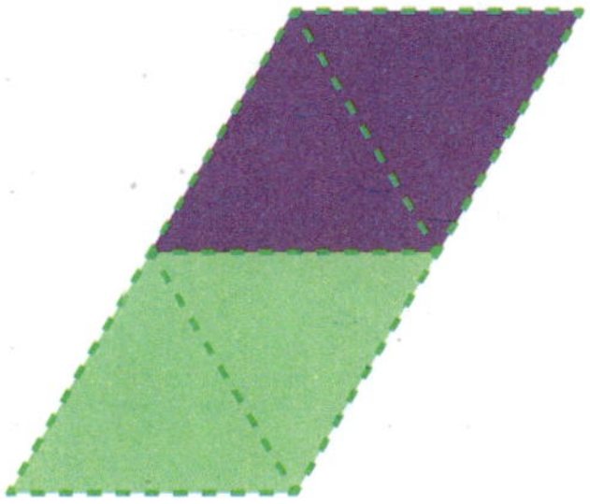 | Color 4 triangles.

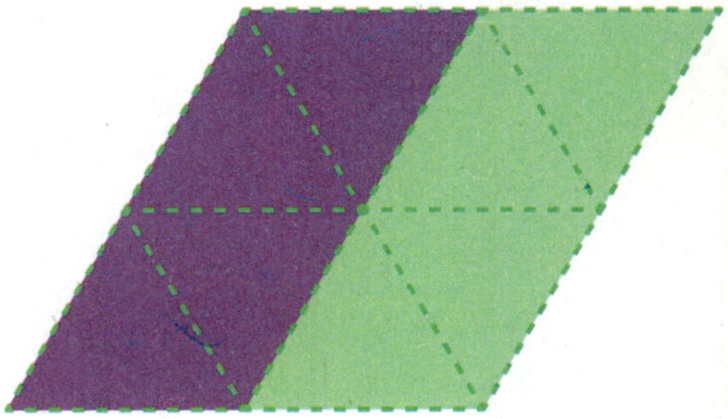

What is the last composed shape?   **rhombus**

It has **4** sides.

I color 1 triangle to make a composed shape.

I color 2 triangles to make a new composed shape.

I color 4 triangles to make a composed shape with 4 sides.

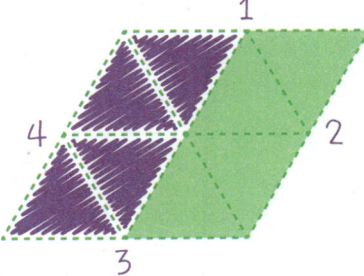

The composed shape is a rhombus.

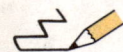

2. Draw a triangle to compose a rectangle.

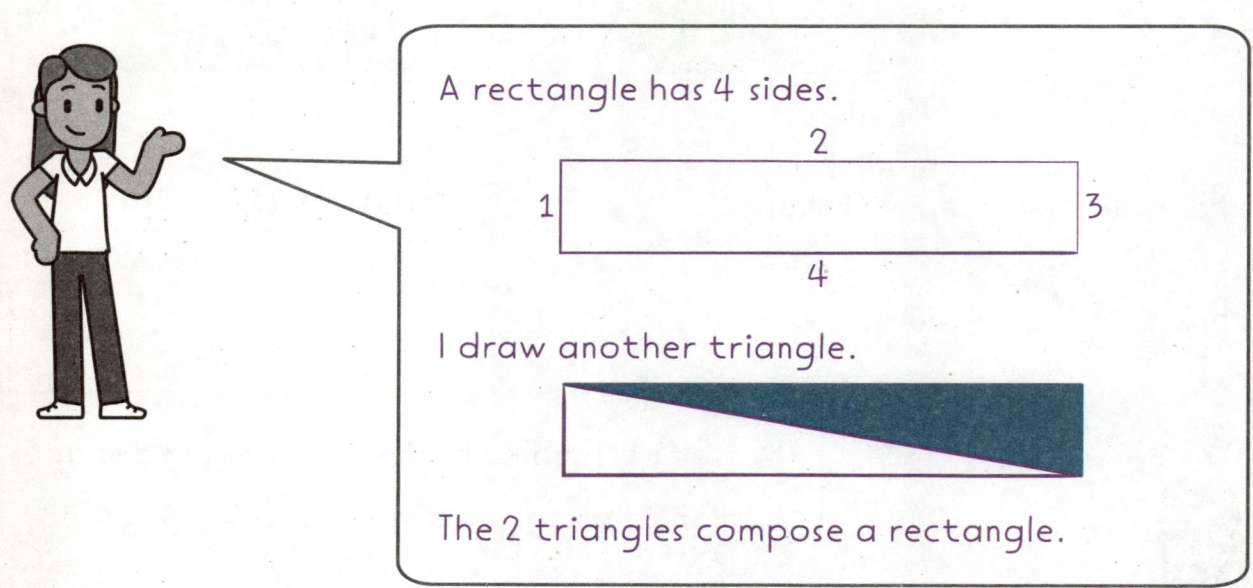

A rectangle has 4 sides.

I draw another triangle.

The 2 triangles compose a rectangle.

Name

1. Color 1 triangle. | Color 2 triangles. | Color 4 triangles.

What is the last new shape? _____

It has  sides.

2. Draw a triangle to compose a rectangle.

3. Draw a triangle to compose a rhombus.

EUREKA MATH²   1 ▸ M6 ▸ TB ▸ Lesson 9

Name

1. Circle the hexagon made with **more** shapes.

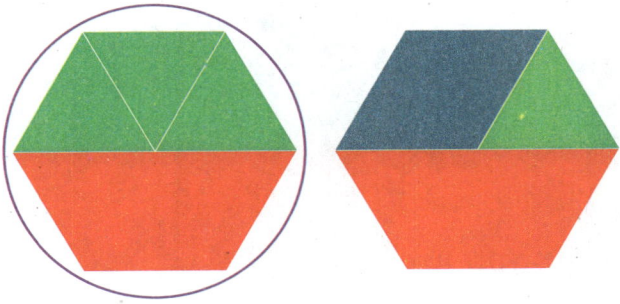

Draw to make a hexagon that has more shapes.

Sample:

The first hexagon has 4 shapes.

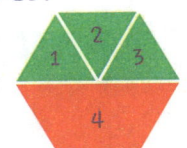

The second hexagon has 3 shapes.

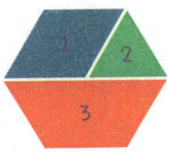

I make my hexagon using more than 4 shapes.

I draw 3 triangles on top.

 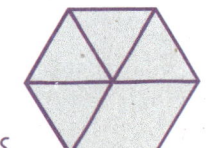

I add a triangle and a rhombus.

2. Draw to make a quadrilateral made of shapes.

   Sample:

   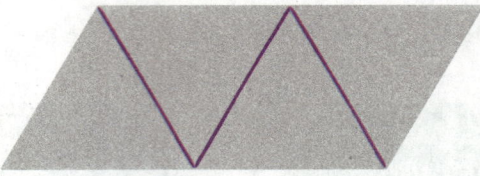

   How many shapes did you use? 4

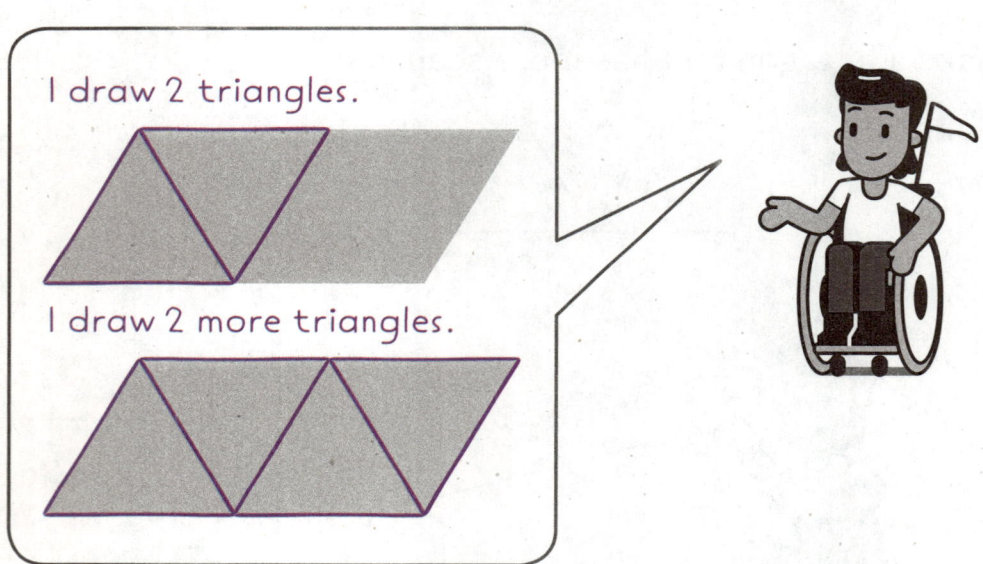

I draw 2 triangles.

I draw 2 more triangles.

REMEMBER

3. **Read**

Sam's mouse is 5 centimeters long.

Jon's mouse is 10 centimeters longer than Sam's mouse.

How long is Jon's mouse?

Draw Sample:

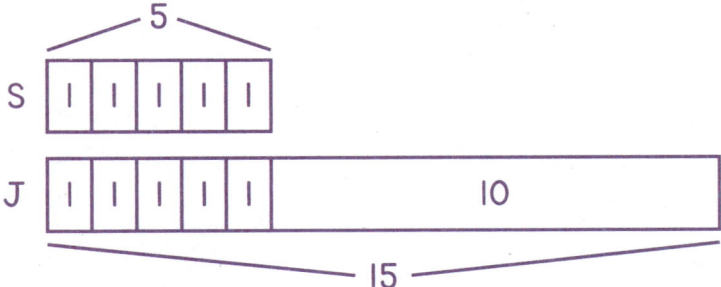

> I read the problem. I read again. As I reread, I think about what I can draw.
>
> I draw 5 centimeter cubes to show the length of Sam's mouse.
>
> I draw 5 centimeter cubes and a 10-stick to show Jon's mouse is 10 centimeters longer than Sam's mouse.
>
> I need to figure out how long Jon's mouse is.
>
> I can add 5 and 10.

Write

5 + 10 = 15

 Jon's mouse is 15 centimeters long.

Name _____

1. Circle the hexagon made with **fewer** shapes.

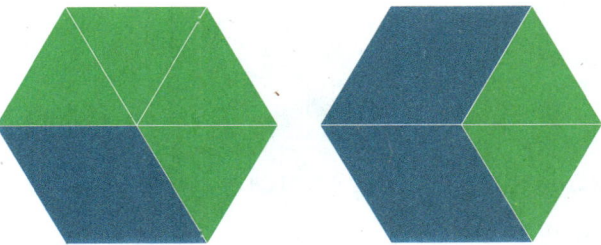

Draw to make a hexagon with fewer shapes.

2. Draw to make a triangle made of shapes.

How many shapes did you use?

REMEMBER

3. **Read**

Kit's lizard is 9 centimeters long.

Ben's lizard is 10 centimeters longer than Kit's lizard.

How long is Ben's lizard?

Draw

Write

Ben's lizard is _____ centimeters long.

Module 6
Topic C

FAMILY MATH
Halves and Fourths

Dear Family,

Your student is exploring the idea of equal shares by folding, cutting, or drawing lines to partition or split shapes. They discuss whether the parts or shares that result are equal and learn to name the equal-sized parts as halves and fourths, or quarters. Then they reason about the number of equal shares and the size of the shares. Students come to understand that the more shares of a whole there are, the smaller each share is. Your student is also applying this understanding to tell time to the half hour using the term *half past*.

Key Terms

fourth/fourths/fourth of

half/halves/half of

half past

partition

quarter/quarter of

The paper is folded into 2 equal parts called halves. 1 part is 1 half of the whole.

The pie is partitioned into 4 equal shares called fourths or quarters. 1 part is 1 fourth of the whole.

1 half of a pizza is larger than 1 fourth of the same size pizza. The shares are smaller when the pizza is cut into more parts.

The clock is in the shape of a circle. The minute hand has gone halfway around the clock, so it is half past, or 30 minutes past the hour.

At-Home Activities

Fold Equal Shares

Practice folding paper into equal shares with your student. Provide a newspaper page, a paper plate, or some other scrap paper, and ask your student to fold it into 2 shares. Discuss whether the shares are equal, and why. Have them name equal shares as halves. Try the activity again by folding an item into 4 equal shares and naming the shares as fourths or quarters.

Playful Partitions

Find opportunities to review halves and fourths with your student in everyday activities. For example, consider doing the following when you are preparing food.

- Slice a sandwich into 2 or 4 equal pieces and ask your student to name each piece as 1 half or 1 fourth.
- Invite your student to break food items such as graham crackers or cheese slices into halves or quarters.
- Discuss where to make cuts in food to partition them into halves or fourths.

EUREKA MATH²   1 ▸ M6 ▸ TC ▸ Lesson 10

10

Name

1. Circle shapes that show equal parts.

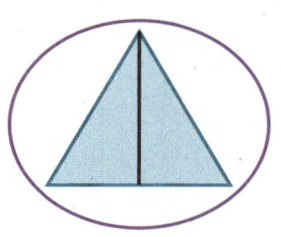 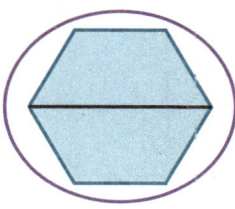 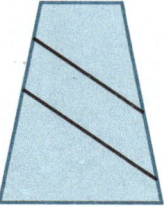

I know the triangle shows equal parts. Both parts are the same size and shape.

I know the hexagon shows equal parts. Both parts are the same size and shape.

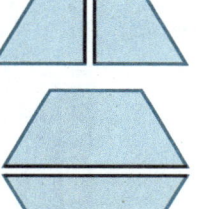
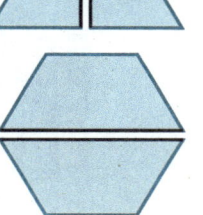

2. Circle foods that show equal shares.

 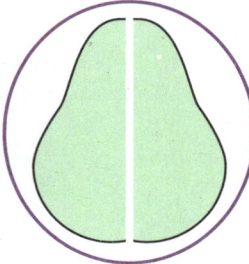 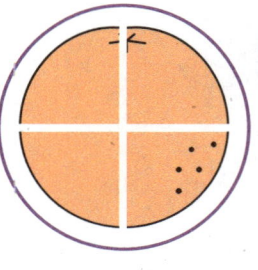

The orange shows equal shares. All 4 parts are the same size and shape.

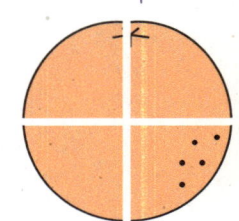

The pear shows equal shares. Both parts are the same size and shape.

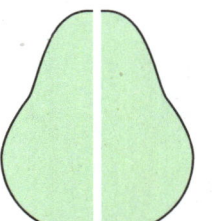

3. Draw lines to make equal parts.

Sample:

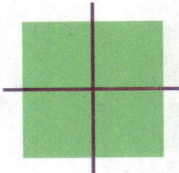 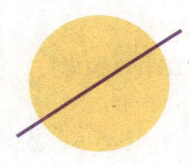 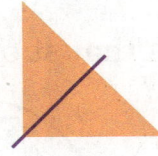

I partition a shape when I draw, cut, or fold it into parts.

I draw lines to **partition** the shapes into equal parts that are the same size and shape.

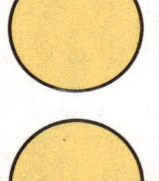

EUREKA MATH²

Name

1. Circle shapes that show equal parts.

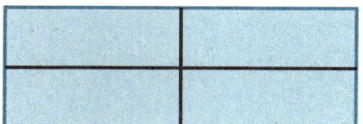   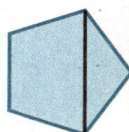

2. Circle foods that show equal shares.

 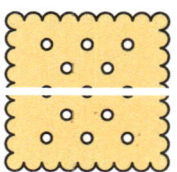  

3. Draw lines to make equal parts.

EUREKA MATH²   1 ▸ M6 ▸ TC ▸ Lesson 11

Name

1. Circle the shapes that show **halves**.

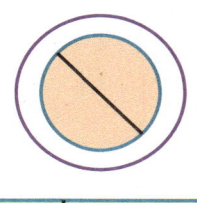
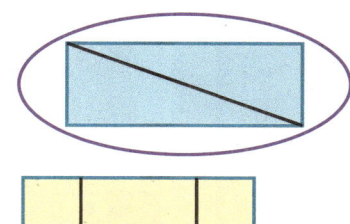
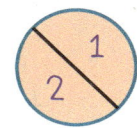
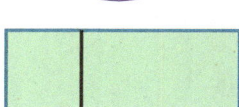
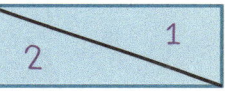

Halves have 2 equal parts.

I see a circle partitioned into halves.

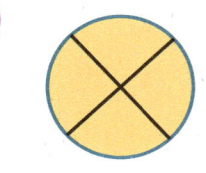

I see a rectangle partitioned into halves.

2. Circle the shapes that show **fourths**.

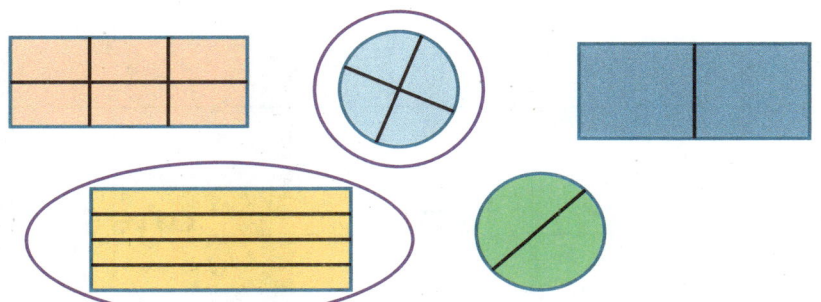

Fourths have 4 equal parts.

I see a circle partitioned into fourths.

I see a rectangle partitioned into fourths.

Copyright © Great Minds PBC   47

3. Draw to make **halves**.   Sample:

   Color 1 **half**.

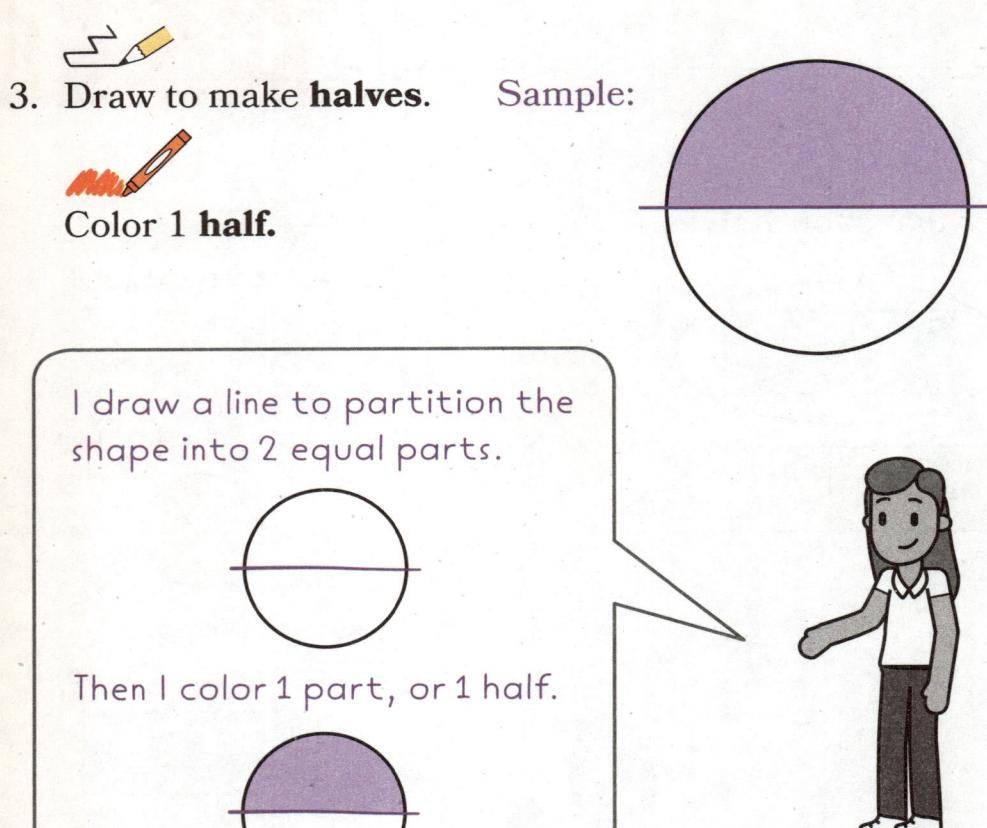

REMEMBER

4. Show ways to make 24.

   Use ones, or tens and ones.

___0___ tens ___24___ ones

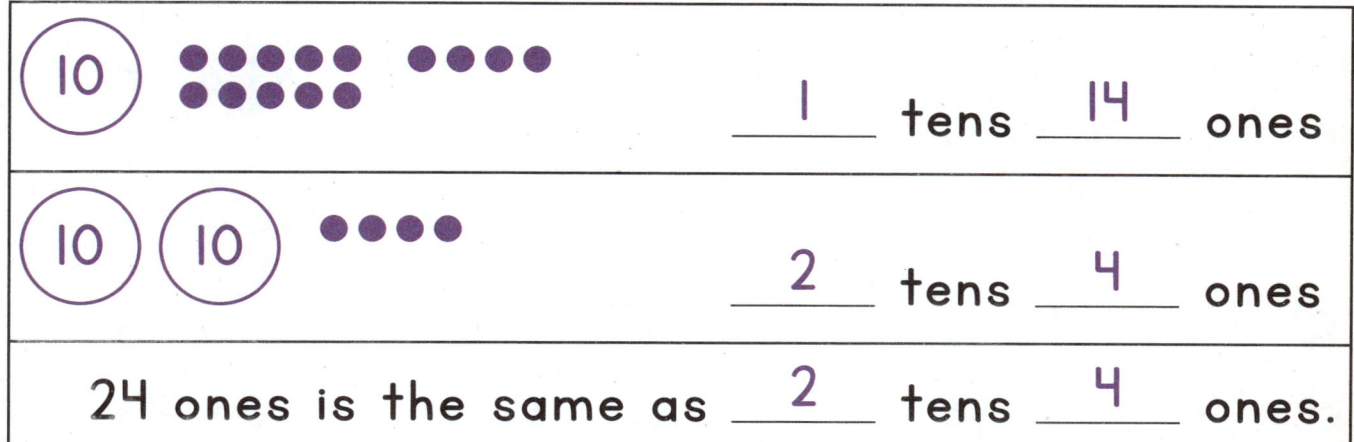

⑩ ●●●●● ●●●● ●●●●	__1__ tens __14__ ones
⑩ ⑩ ●●●●	__2__ tens __4__ ones
24 ones is the same as __2__ tens __4__ ones.	

I can show the number 24 in different ways.
I know that 24 is the same as 24 ones.

I can trade 10 ones for 1 ten. 24 is the same as 1 ten 14 ones.

I can trade 10 more ones for another ten. 24 is the same as 2 tens 4 ones.

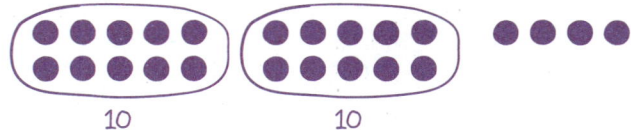

Name _____

1. Circle the shapes that show **halves**.

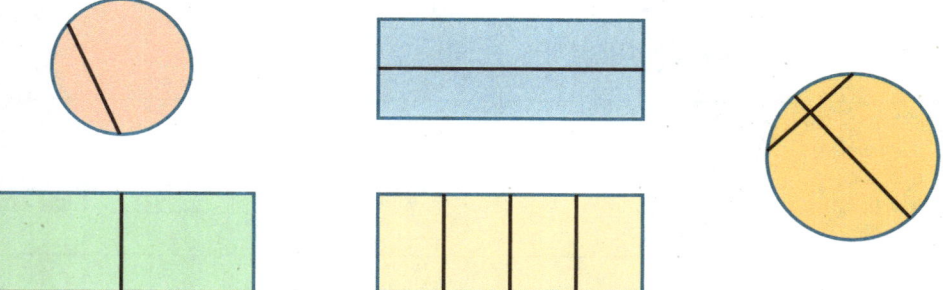

2. Circle the shapes that show **fourths**.

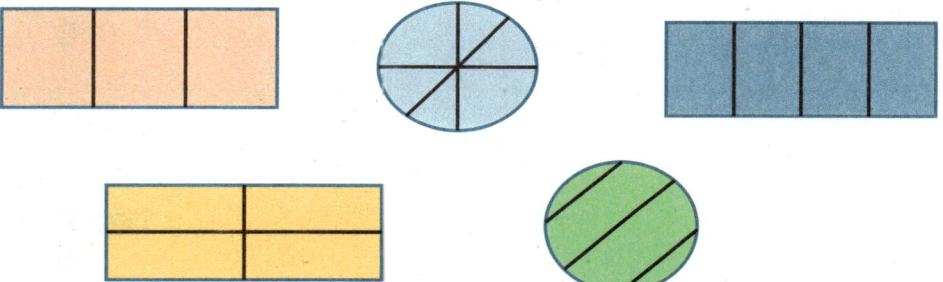

3. Draw to make **fourths**.

Color 1 **fourth**.

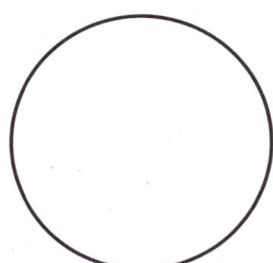

REMEMBER

4. Show ways to make 35.

   Use ones, or tens and ones.

_____ tens _____ ones
_____ tens _____ ones
_____ tens _____ ones
_____ tens _____ ones
35 ones is the same as _____ tens _____ ones.

EUREKA MATH²   1 ▸ M6 ▸ TC ▸ Lesson 12

12

Name

1. Circle.

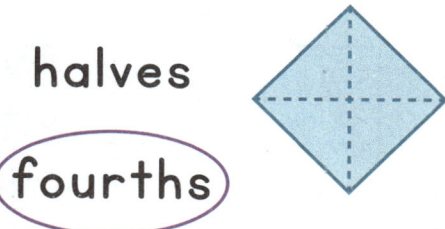

How many equal parts?

4

2. Circle.

How many equal parts?

2

I know that equal parts are the same size and shape.

Halves are 2 equal parts.

Fourths are 4 equal parts. Another word for fourths is **quarters**.

3. Draw a line to make **halves**.   Draw lines to make **quarters**.

   Color 1 **half**.   Color 1 **quarter**.

I partition the rectangle into 2 equal parts to show halves.

I color 1 part, or 1 half.

I partition the rectangle into 4 equal parts to show quarters.

I color 1 part, or 1 quarter.

Name _____

1. Circle.

halves

fourths

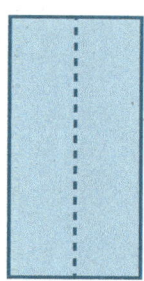

How many equal parts?

2. Circle.

halves

quarters

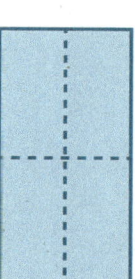

How many equal parts?

3.  Draw a line to make **halves**.

    Color 1 **half**.

   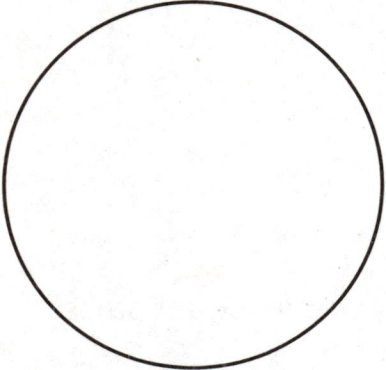

    Draw lines to make **quarters**.

    Color 1 **quarter**.

   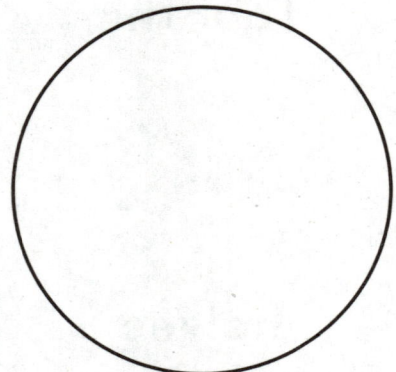

EUREKA MATH²   1 ▸ M6 ▸ TC ▸ Lesson 13

Name

1. Color 1 share.

Circle the shape with the **smaller** shares.

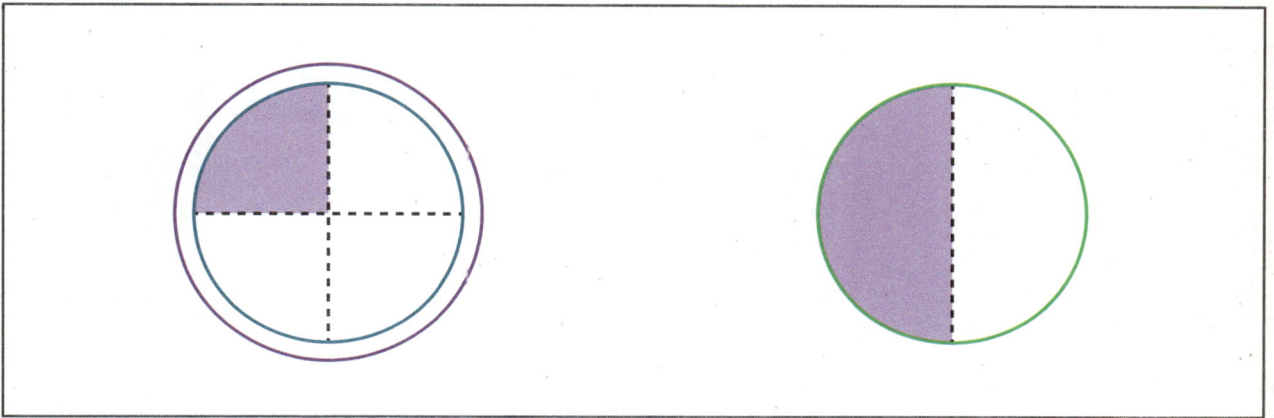

I color a fourth and a half.
I know shapes partitioned into more shares or parts have smaller shares.

4 shares   2 shares

The shape partitioned into fourths has the smaller shares.

2. Color 1 share.

   Circle the shape with the **larger** shares.

   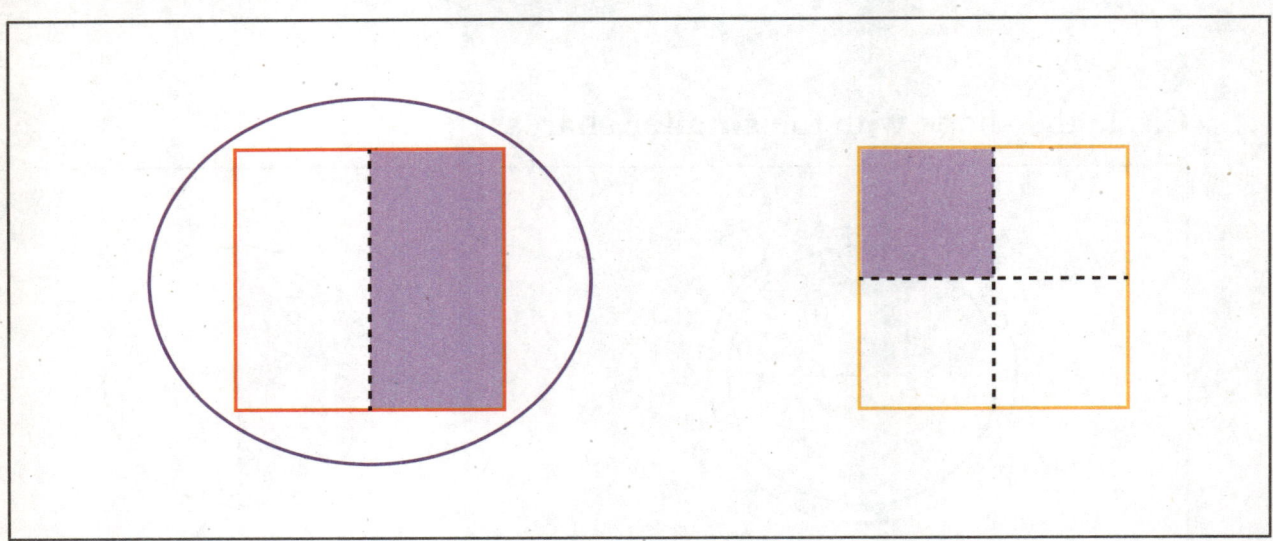

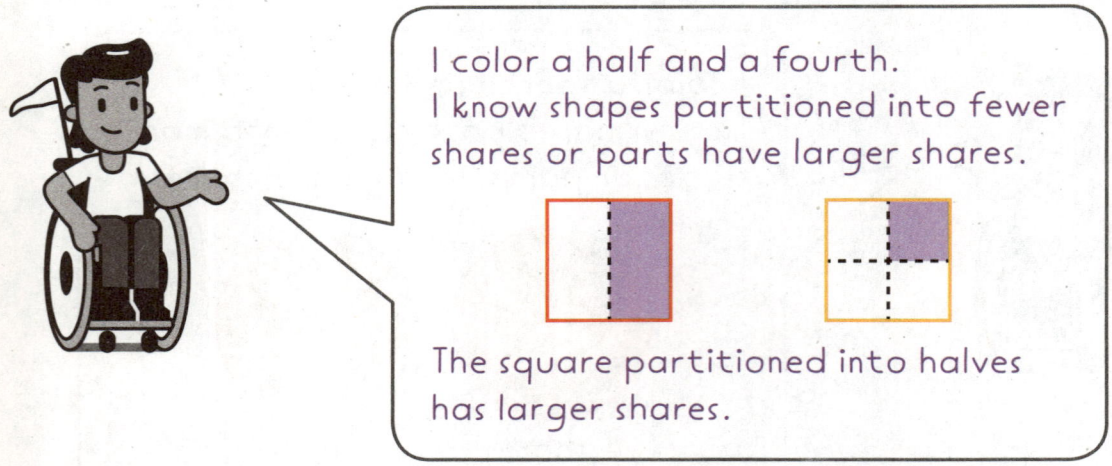

I color a half and a fourth.
I know shapes partitioned into fewer shares or parts have larger shares.

The square partitioned into halves has larger shares.

REMEMBER

3. Draw tens to subtract.

70 − 30 = **40**

__7__ tens − __3__ tens = __4__ tens

> I know the tens place tells me how many tens I need to draw.
>
> There are 7 tens in 70. I draw 7 quick tens.
>
>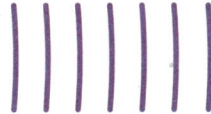
>
> There are 3 tens in 30. I cross off 3 tens to subtract 30.
>
>
>
> 70 − 30 = 40
>
> 4 tens is 40.

Name _____

1. Color 1 share.

Circle the shape with the **smaller** shares.

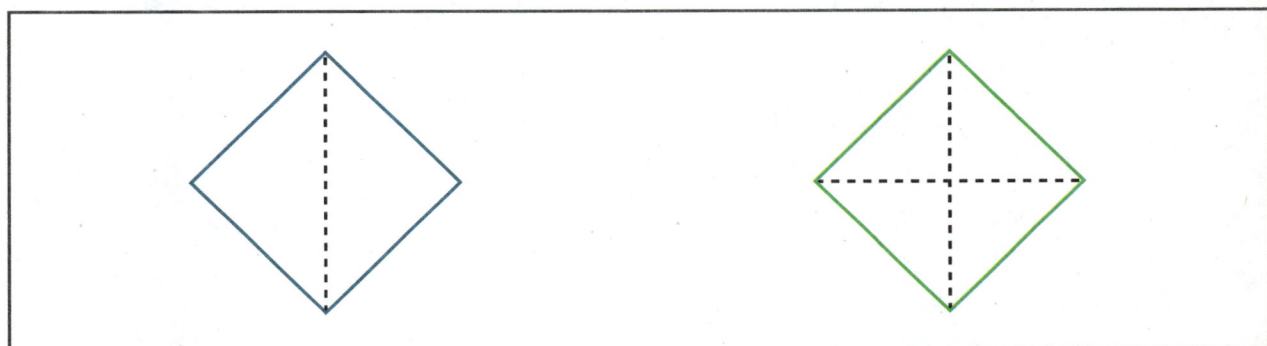

2. Color 1 share.

Circle the shape with the **larger** shares.

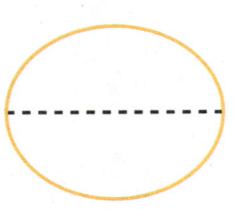 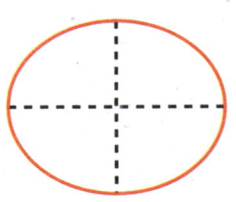  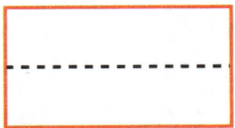

REMEMBER

3. Draw tens to subtract.

90 − 20 = ☐

____ tens − ____ tens = ____ tens

EUREKA MATH²   1 ▸ M6 ▸ TC ▸ Lesson 14

Name

1. Circle all clocks that show half past 8.

This clock shows 30 minutes past the hour.

Half past 8 is the same as 8:30.

This clock shows 8:00.

This clock shows 8:30.

2. Draw lines to match the times.

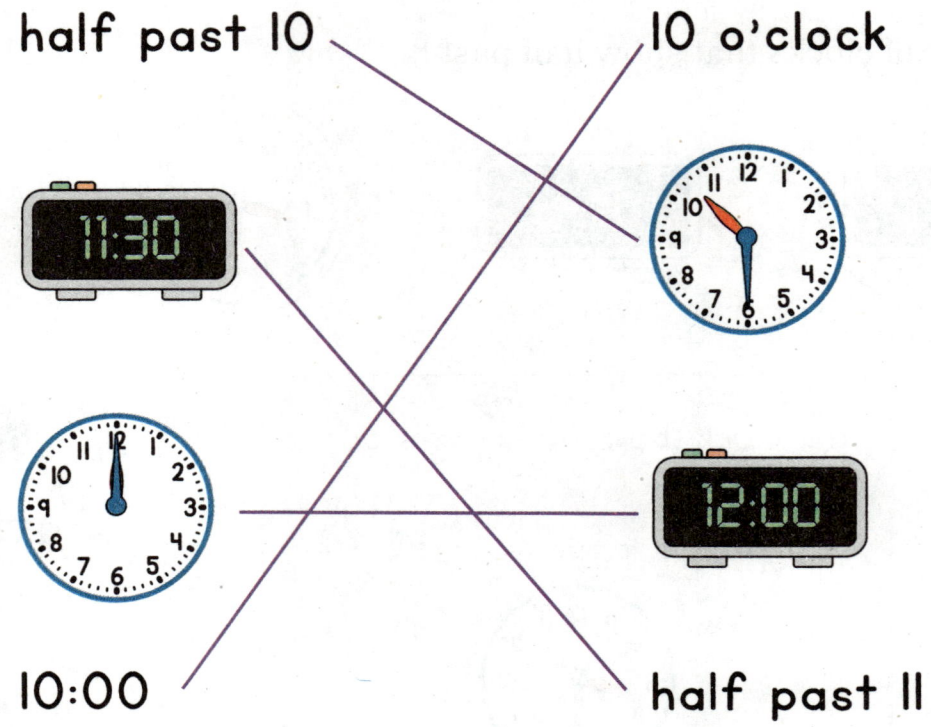

EUREKA MATH²   1 ▸ M6 ▸ TC ▸ Lesson 14

14

Name

1. Circle all clocks that show half past 7.

2. Draw lines to match the times.

1 o'clock

1:30

half past 4

4 o'clock

half past 1

2:00

2:30

EUREKA MATH²   1 ▸ M6 ▸ TC ▸ Lesson 15

15

Name

1. Draw or write to show 9 o'clock.

   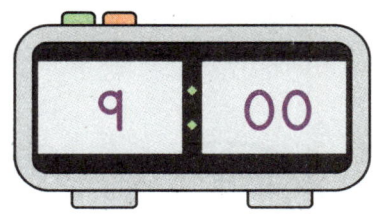

2. Draw or write to show half past 9.

I know that 9 o'clock is the same as 9:00.

The minute hand points to 12 when it is 9:00.

I know that half past 9 is the same as 9:30.

The minute hand points to 6 when it is 9:30.

3. Circle all clocks that show half past 5.

I know that half past 5 is the same as 5:30.

The hour hand is between 5 and 6 at half past 5.

The minute hand points to 6 when it is half past 5.

REMEMBER

4. Write <, >, or =.

   45 < 54

5. Circle the shapes with parallel sides.

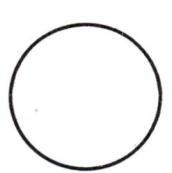 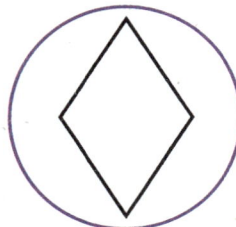 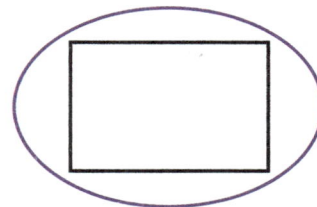

> I can draw tens and ones to compare the numbers.
> 45 has 4 tens 5 ones. 54 has 5 tens 4 ones.
>
>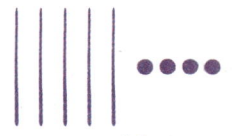
>
> I look at the tens place first to compare 45 and 54.
> 4 tens is less than 5 tens.
> 45 is less than 54.

> I know **parallel** sides never touch and are across from each other.
> I can use craft sticks or other straight objects to check for parallel sides.
> The sticks on the rectangle and rhombus will not touch.
>
>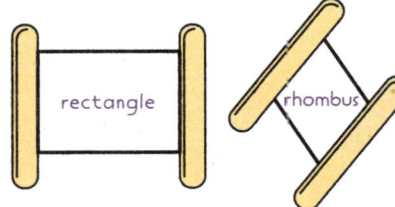
>
> The circle does not have straight sides.
> The rectangle and the rhombus have parallel sides.

Name

1. Draw or write to show 10 o'clock.

2. Draw or write to show half past 10.

3. Circle all clocks that show half past 7.

REMEMBER

4. Write <, >, or =.

   28 82

5. Circle the shapes with parallel sides.

 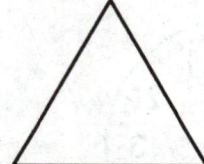 

Module 6
Topic D

FAMILY MATH
Count and Represent Numbers Beyond 100

Dear Family,

Your student is extending their understanding of numbers as they now work with three-digit numbers from 100 to 120. They pay careful attention when counting up and down across a ten, such as 99, 100, 101 or 111, 110, 109. They count by ones or by tens, and they identify unknown numbers in a sequence. Your student learns to count collections of more than 100 objects that are arranged in groups of tens and ones. They represent the totals in various combinations of hundreds, tens, and ones. This work builds the understanding that 10 tens is 100, which helps your student to read and write numbers greater than 100.

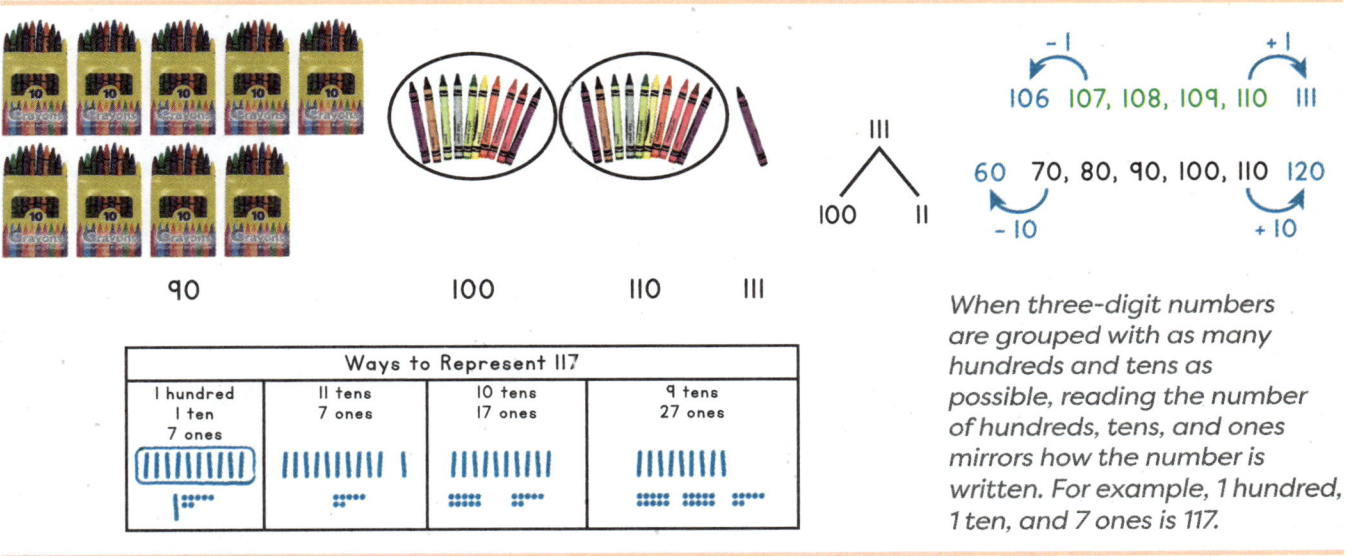

When three-digit numbers are grouped with as many hundreds and tens as possible, reading the number of hundreds, tens, and ones mirrors how the number is written. For example, 1 hundred, 1 ten, and 7 ones is 117.

At-Home Activities

Count Up and Down Across 100

Write the number 100 on a sheet of paper. Ask your student to choose a number between 1 and 6 and draw that number of blanks on either side of the 100. Then have your student count up and count down from 100 by ones to fill in the blanks. For example, if your student chooses 3, draw 3 blanks on either side of 100 (_____, _____, _____, 100, _____, _____, _____), and have your student fill in the blanks (97, 98, 99, 100, 101, 102, 103). For more practice, consider repeating the activity starting with numbers between 95 and 110. As a challenge, have your student count by tens to fill in the blanks, helping them with numbers greater than 120.

Ways to Represent Numbers Greater than 100

Give your student a book that has at least 120 pages. Ask them to turn to a page located between page 100 and page 120 in the book and record the page number. Then have your student make three place value drawings to represent the number in three different ways. They may draw lines for tens and dots for ones.

EUREKA MATH²   1 ▸ M6 ▸ TD ▸ Lesson 17

Name

1. Write the part or the total.

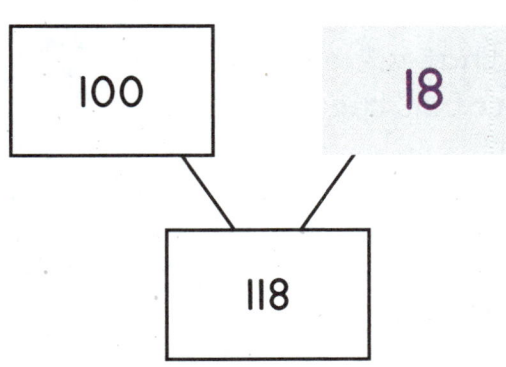

Total 104

The total is 118.

I know 118 can be broken up into 100 and some more.

100 + 18 = 118

I see 10 trees with 10 apples on each tree.

I know 10 tens is 100.

There are 4 extra apples.

100 and 4 is 104.

REMEMBER

2. Subtract.

   10 − 10 = **0**

   10 − 9 = **1**

   10 = 10 − 0

   10 − 1 = **9**

When you subtract all, you get 0.

When you subtract a number that is 1 less than the total, the answer is 1.

When you subtract 0, you get the number you started with.

When you subtract 1, the answer is the number that comes right before.

3. Color the triangle.

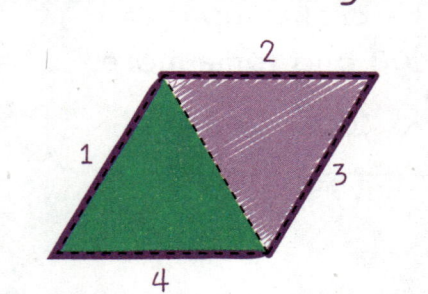

When I color the triangle, I make a composed shape.

The new shape has 4 sides that are the same length.

Trace the new shape. It has **4** sides.

What is the composed shape? **rhombus**

Name

1. Write the part or the total.

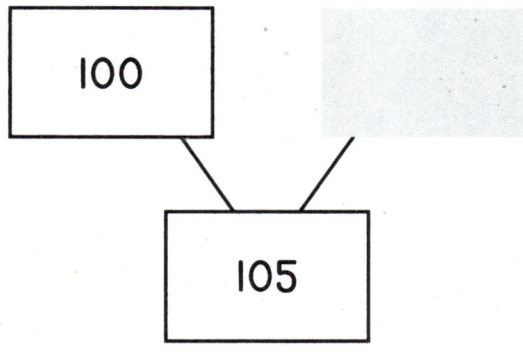

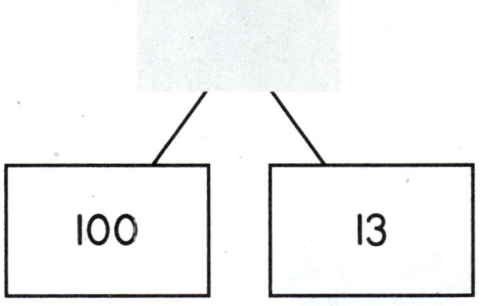

Total

REMEMBER

2. Subtract.

6 − 6 = ☐	8 − 7 = ☐
 = 20 − 0	15 − 1 = ☐

3. Color the triangles.

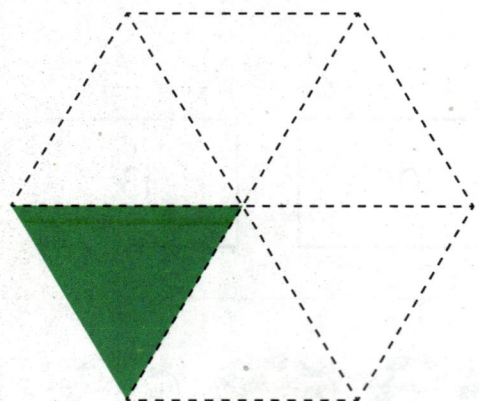

Trace the new shape. It has  sides.

What is the new shape? _____

EUREKA MATH² 1 ▸ M6 ▸ TD ▸ Lesson 18

Name _____

1. Count up.

   Write the numbers.

   84, 85, 86, __87__, __88__, __89__, __90__, __91__

2. Count down.

   Write the numbers.

   __32__, __42__, __52__, __62__, __72__, 82, 92, 102

The numbers are in order.
The pattern tells me how to count.
I see the numbers go up by 1 each time.
I count up by ones to finish the pattern.

+1 +1 +1 +1 +1 +1 +1
84, 85, 86, 87, 88, 89, 90, 91

The numbers go down by 10 each time. I count down, or back, by tens to finish the pattern.

−10 −10 −10 −10 −10 −10 −10
32, 42, 52, 62, 72, 82, 92, 102

79

3. Count by ones.

   Write the numbers.

   __79__, __80__, 81, __82__, __83__

4. Count by tens.

   Write the numbers.

   __54__, __64__, 74, __84__, __94__

> I count down by ones to find the numbers before 81.
>
> I count up by ones to find the numbers after 81.
>
>   −1 −1 +1 +1
> 79, 80, 81, 82, 83

> I can skip-count by tens to find the unknown numbers.
>
> The ones place does not change. The tens place goes up or down by 1 ten.
>
> I count down by tens to find the numbers before 74.
>
> I count up by tens to find the numbers after 74.
>
>   −10 −10 +10 +10
> 54, 64, 74, 84, 94

EUREKA MATH²   1 ▸ M6 ▸ TD ▸ Lesson 18

Name _____

1. Count up.

   Write the numbers.

   76, 77, 78, _____, _____, _____, _____, _____

2. Count down.

   Write the numbers.

   _____, _____, _____, _____, _____, 99, 109, 119

3. Count by ones.

   Write the numbers.

   _____, _____, 105, _____, _____

4. Count by tens.

   Write the numbers.

   _____, _____, 93, _____, _____

EUREKA MATH²   1 ▸ M6 ▸ TD ▸ Lesson 19

19

Name

1. Write the total.

Show how you know.   Sample:

9 tens 16 ones is the same as __106__.

90

100

106

I draw 9 tens and 16 ones.

I can compose a 10 with the 10 ones. Now I have 10 tens.

10 tens is 100.

100 and 6 more is 106.

2. Write a number greater than 99.

Show it in more than one way.   Sample:

$\underline{112}$

10 tens
12 ones

| | | | | | | | | | 90
• • • • • • • • • •   100
• • • • • • • • • •   110
• •   112

The number 112 is greater than 99. I know this because 112 is 100 and some more.

One way I can show 112 is with 10 tens 12 ones.

Another way I can show 112 is by drawing 9 tens first to make 90.

Then, I can count up from 90 to 112 by ones. I draw dots as I count.

I made 9 tens 22 ones.

REMEMBER

3. **Read**

   9 people are on a bus.

   Some more people get on.

   Now there are 12 people on the bus.

   How many people got on the bus?

   Draw Sample:

   > I read the problem. I read again. As I reread, I think about what I can draw.
   >
   > I draw 9 dots for the people on the bus.
   >
   > I draw more dots and count on to 12.
   >
   > I circle the 3 dots I counted on.
   >
   > I write an addition number sentence to show I counted on.

   Write

   $9 + 3 = 12$

   3 people got on the bus.

Name _____

1. Write the total.

Show how you know.

9 tens 21 ones is the same as _____.

2. Write a number greater than 99.

Show it in more than one way.

REMEMBER

3. **Read**

   6 people are on a bus.

   Some more people get on.

   Now there are 11 people on the bus.

   How many people got on the bus?

   Draw

   Write

   _____ people got on the bus.

FAMILY MATH
Deepening Problem Solving

Module 6
Topic E

Dear Family,

Your student is continuing to learn how to solve different types of addition and subtraction word problems by using the familiar Read–Draw–Write process. Before learning this process, your student represented word problems with drawings. Now your student is learning to draw a tape diagram to represent the problem. After drawing a tape diagram, your student is writing an addition or subtraction equation to match it and then solving the problem. Your student is also using small objects such as paper clips and cubes to measure length and connecting this activity to solving comparison word problems.

Key Term
tape diagram

Max has 12 tickets. Kit has 8 tickets. They need 20 tickets to go on the ride. Can they ride?	Jade has 12 tickets. She gets more tickets to go on a ride. Now she has 20 tickets. How many tickets did she get?	Kit plays for 10 minutes. Ben plays for 4 fewer minutes than Kit plays. How long does Ben play?
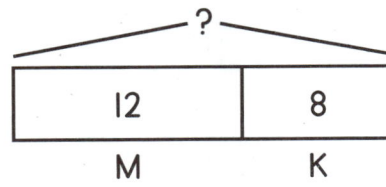	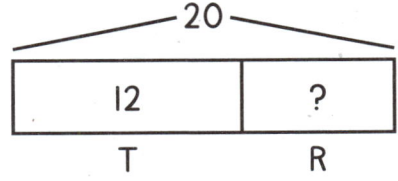	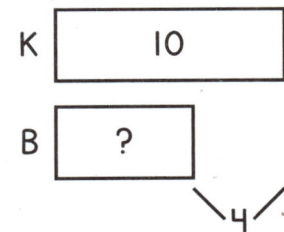
12 + 8 = ? Yes, they can ride. They have 20 tickets.	12 + ? = 20 Jade got 8 more tickets.	10 − 4 = ? Ben plays for 6 minutes.

A tape diagram is a flexible way to show what is known and unknown in a word problem throughout elementary school.

At-Home Activities

Best Foot Forward

Use small items, such as crayons or cotton swabs, to measure the length of family members' feet. Remind your student to line up the objects from end to end without any gaps or overlaps. For example, your student's foot may be about 3 crayons long. Encourage your student to compare lengths using words such as *longer* and *shorter*. Your student may say, "My foot is 2 crayons shorter than your foot." Consider measuring the feet again with another object and discussing how the measurements are different.

My World Puzzlers

Look for opportunities to have your student solve addition and subtraction problems at home. The numbers in the problems should be 20 or less. Encourage your student to draw a tape diagram and write an equation to figure out the answer. If possible, check their answers with the actual objects. Here are a few examples.

- We baked 20 cookies. Then we ate some of the cookies. Now there are 11 cookies left. How many cookies did we eat?
- I need 13 stamps to mail these cards. I found 8 stamps on the table and 4 stamps in the drawer. Do I have enough stamps to mail all of the cards?
- The drive to the lake takes 4 hours. That trip takes 2 fewer hours than the drive to the beach. How long is the drive to the beach?

Name

1. **Read**

    Deb has 13 stamps.

    She gets some more stamps.

    Now Deb has 18 stamps.

    How many more stamps did Deb get?

    Draw Sample:

    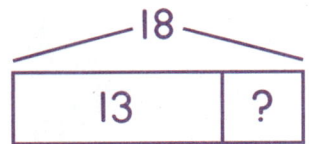

> I read the problem. I read again. As I reread, I think about what I can draw.
>
> I draw a tape diagram.
>
> Deb starts with 13 stamps and then gets some so I write 13 in one part.
>
> I don't know how many more she gets.
>
> Deb has 18 now so I write 18 as the total.
>
> I need to figure out how many more stamps Deb got.
>
> I can count on from 13 to 18 to find the unknown part.

Write

$$13 + 5 = 18$$

Deb got 5 more stamps.

2. **Read**

Ben has 17 stamps.

He uses some stamps.

Ben has 14 stamps left.

How many stamps did Ben use?

Draw Sample:

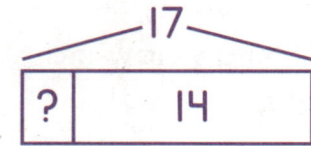

I read the problem. I read again. As I reread, I think about what I can draw.

I draw a tape diagram.

Ben starts with 17 stamps and uses some so I write 17 as the total.

I do not know how many he uses.

He has 14 stamps left so I write 14 in one part.

I need to figure out how many stamps Ben uses.

I can count back from 17 to 14 to find the unknown part.

Write

17 − 14 = 3

Ben used   3   stamps.

REMEMBER

3. Make 10 to add.

   Show how you know.

   6 + 5 + 4 = **15**

   I can think of partners to 10 to make adding easier.

   I know 6 and 4 make 10, so I add them first.

   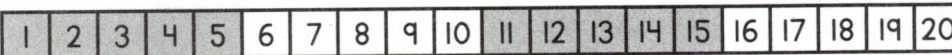

   Next, I add 10 and 5. 10 and 5 make 15.

4. Count back to 10 to subtract.

   14 − 6 = **8**

   I start at 14 and hop back to 10. I label my hop.

   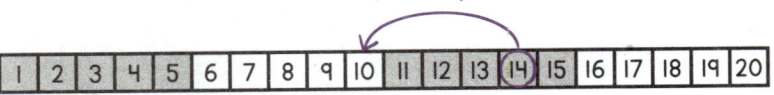

   I need to subtract 6. I know 4 and 2 make 6 so I hop back 2 more.

   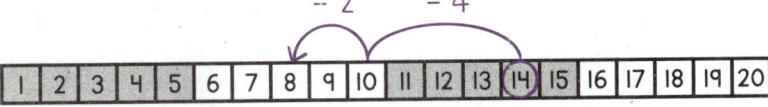

   I land on 8.

Name _____

1. **Read**

   Baz has 7 rocks.

   He finds more rocks.

   Now Baz has 11 rocks.

   How many more rocks did Baz find?

   Draw

   Write

   Baz found _____ more rocks.

2. **Read**

   Beth has 15 shells.

   She gives some away.

   Beth has 9 shells left.

   How many shells did Beth give away?

   Draw

   Write

   Beth gave away _____ shells.

REMEMBER

3. Make 10 to add.

    Show how you know.

   $$7 + 4 + 3 = \boxed{}$$

4. Count back to 10 to subtract.

   $$16 - 9 = \boxed{}$$

   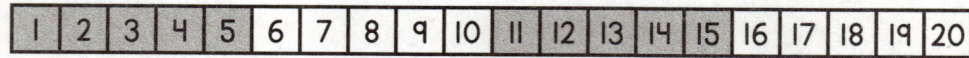

Name _____

22

1. **Read**

   Hope has some crackers.

   She gets 7 more.

   Now Hope has 16 crackers.

   How many crackers did Hope have to start?

   Draw Sample:

   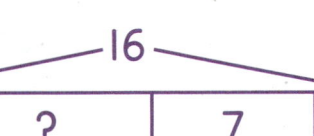

   I read the problem. I read again. I think about what I can draw.

   I draw a tape diagram.

   I do not know how many crackers Hope starts with.

   She gets 7 more so I write 7 in one part.

   Now she has 16 so I write 16 as the total.

   I need to find the part Hope started with.

   I can subtract 7 from 16 to find the unknown part.

   Write

   16 − 7 = 9

   Hope had   9   crackers.

97

2. **Read**

Jon has some cookies.

He gives 6 away.

Jon has 7 cookies left.

How many cookies did Jon have to start?

Draw Sample:

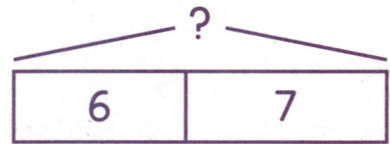

> I read the problem. I read again. I think about what I can draw.
>
> I draw a tape diagram.
>
> I do not know how many cookies he starts with.
>
> Jon gives 6 cookies away, so I write 6 in one part.
>
> He has 7 cookies left so I write 7 in the other part.
>
> I need to figure out how many total cookies Jon started with.
>
> I can add the parts, 6 and 7, to find the total.

Write

6 + 7 = 13

Jon had 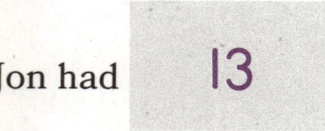 13  cookies.

Name _____

1. **Read**

   Dan has some grapes.

   He gets 5 more grapes.

   Now Dan has 12 grapes.

   How many grapes did Dan have to start?

   Draw

   Write

   Dan had _____ grapes.

2. **Read**

   Jade has some pencils.

   She gives 4 pencils to friends.

   Jade has 8 pencils left.

   How many pencils did Jade have to start?

   Draw

   Write

   Jade had _____ pencils.

EUREKA MATH² 　　　1 ▸ M6 ▸ TE ▸ Lesson 23

Name _____

1. **Read**

   Peg reads for 12 minutes.

   Zan reads for 9 minutes.

   How many more minutes does Peg read than Zan reads?

   Draw Sample:

   P | 12
   Z | 9
   ?

   I read the problem. I read again. I think about what I can draw.

   I draw and label a tape diagram to compare the minutes.

   I draw to show Peg read for 12 minutes.

   I draw below Peg to show Zan read for 9 minutes.

   I need to figure out how many more minutes Peg reads. I can count on from 9 to 12.

   Write

   $9 + 3 = 12$

   Peg reads for ⬚3⬚ more minutes.

2. **Read**

Sam reads for 11 minutes.

Mel reads for 7 fewer minutes than Sam reads.

How many minutes does Mel read?

Draw Sample:

```
S |     11     |
M | ? |
       \_7_/
```

I read the problem. I read again. I think about what I can draw.

I draw and label a tape diagram to compare the minutes.

I draw to show Sam reads for 11 minutes.

I draw below Sam to show that Mel reads for fewer minutes.

I write to show that Mel reads for 7 fewer minutes than Sam.

I need to figure out how many minutes Mel reads. I can subtract 7 from 11.

Write

11 − 7 = 4

Mel reads for 4 minutes.

REMEMBER

3. Subtract.

    10 − 7 = 3

    > I can think of 10 − 7 as 7 + ___ = 10.
    > I know 7 + 3 = 10.
    > So, 10 − 7 = 3.

    Write the addition sentence that helps you.

    7 + 3 = 10

4. Circle.

    > The popsicle is taller than the ice cream cone.
    > The lollipop is shorter than the ice cream.
    > So, I know the lollipop is shorter than the popsicle.

    The lollipop is (**shorter**) / taller than the popsicle.

    Draw or write to put the ice cream, lollipop, and popsicle in order from shortest to tallest.

      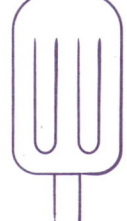

Name

1. **Read**

   Max reads for 10 minutes.

   Nate reads for 7 minutes.

   How many more minutes does Max read than Nate reads?

   Draw

   Write

   Max reads for _____ more minutes.

2. **Read**

   Kit reads for 11 minutes.

   Liv reads for 6 fewer minutes than Kit.

   How many minutes does Liv read?

   Draw

   Write

   Liv reads for _____ minutes.

REMEMBER

3. Subtract.

   $9 - 5 = $

   Write the addition fact that helps you.

4. Circle.

   Draw or write to put the tree, giraffe, and hippo in order from shortest to tallest.

Name _____

1. **Read**

    Deb's kite is 7 crayons long.

    Deb's kite is 4 crayons shorter than Beth's kite.

    How long is Beth's kite?

    Draw Sample:

    ```
         ⌒4⌒
    D [  7  ]
    B [    ?    ]
    ```

    > I read the problem. I read again. I think about what I can draw.
    >
    > I draw and label a tape diagram to compare the kites.
    >
    > I draw to show Deb's kite is 7 crayons long.
    >
    > I draw below Deb to show that Beth's kite is longer.
    >
    > I write to show that Deb's kite is 4 crayons shorter.
    >
    > I need to figure out how long Beth's kite is. I can add to find the length of her kite.

    Write

    $7 + 4 = 11$

    Beth's kite is 11  crayons long.

2. **Read**

Deb's kite is 7 crayons long.

Deb's kite is 4 crayons longer than Hope's kite.

How long is Hope's kite?

Draw Sample:

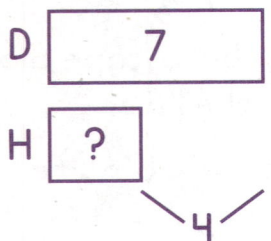

> I read the problem. I read again. I think about what I can draw.
>
> I draw and label a tape diagram to compare the kites.
>
> I draw to show Deb's kite is 7 crayons long.
>
> I draw below Deb to show that Hope's kite is shorter.
>
> I write to show that Deb's kite is 4 crayons longer.
>
> I need to figure out how long Hope's kite is. I can subtract to find the length of her kite.

Write

$7 - 4 = 3$

Hope's kite is 3 crayons long.

Name _____

1. **Read**

   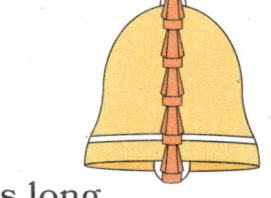

   Ren's bell is 5 erasers long.

   Ren's bell is 3 erasers shorter than Nate's bell.

   How long is Nate's bell?

   Draw

   Write

   Nate's bell is _____ erasers long.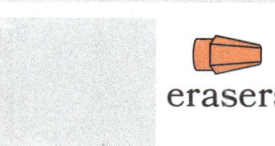

2. **Read**

   Ren's bell is 5 erasers long.

   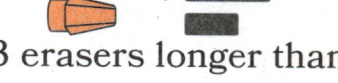

   Ren's bell is 3 erasers longer than Jon's bell.

   How long is Jon's bell?

   Draw

   Write

   Jon's bell is _____ erasers long.

Name _____

1. **Read**

   How many bears do they see?

   How many penguins do they see?

   Draw Sample:

   heads

   $2 + 2 + 4 + 4 = 12$
   legs

   > I read the problem. I read again. I think about what I can draw.
   >
   > I draw 4 circles to show 4 heads.
   >
   > I know there are 12 legs total. Bears have 4 legs, and penguins have 2 legs.
   >
   > On the first 2 heads, I draw 2 legs. That makes 4 legs.
   >
   > I draw 4 legs on the last 2 heads to make 8 more legs.
   >
   > I add the legs and count the heads.

   Write Sample:

   $2 + 2 + 4 + 4 = 12$

   They see   2   bears and   2   penguins.

REMEMBER

2. **Read**

   Mr. West has 18 pencils.

   He loses some pencils.

   He has 11 pencils left.

   How many pencils did Mr. West lose?

   Draw Sample:

   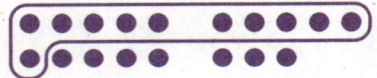

   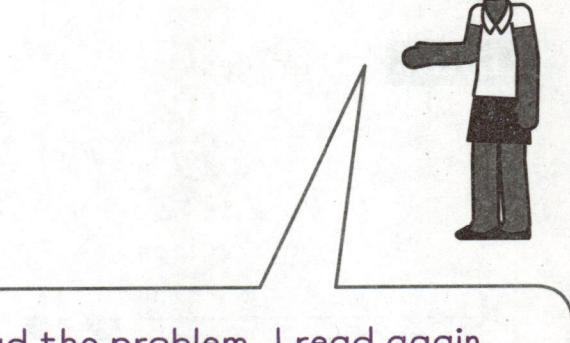

   I read the problem. I read again. I think about what I can draw.

   I draw 18 dots to show Mr. West's pencils.

   I circle the 11 he has left.

   I count how many he lost.

   Write

   $18 - 11 = 7$

   Mr. West lost 7 pencils.

EUREKA MATH² 　　　　　　　　　　　　　　1 ▸ M6 ▸ TE ▸ Lesson 25

25

Name

1. **Read**

   How many goats do they see?

   How many turkeys do they see?

   Draw

Write

They see  goats and  turkeys.

REMEMBER

2. **Read**

There are 19 bunnies.

Some hop away.

Now there are 13 bunnies.

How many bunnies hopped away?

Draw

Write

 bunnies hopped away.

FAMILY MATH
Extending Addition to 100

Module 6
Topic F

Dear Family,

Your student is using familiar drawings and strategies to add two-digit numbers. Now the totals are larger and your student is practicing adding up to 100. They may draw to represent each number by using lines for tens and dots for ones. Then your student chooses how to combine the numbers. Using various strategies, shown here, allows your student to think flexibly and prepares them for adding with larger numbers in later grades.

54 + 28

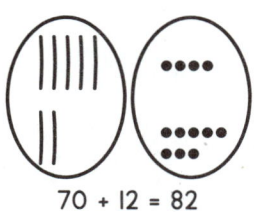

70 + 12 = 82

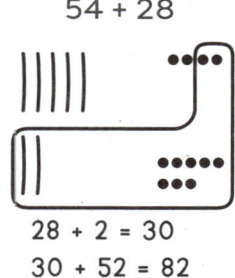

28 + 2 = 30
30 + 52 = 82

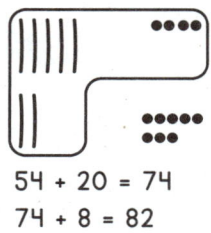

54 + 20 = 74
74 + 8 = 82

Add by grouping all of the tens and all of the ones. This is called adding tens with tens and ones with ones.

28 is close to 30. Adding 2 ones from 54 to the 28 makes this an easier problem: 30 + 52. This is called making the next ten.

Count on from 54 by tens. Then add the remaining ones. This is called adding tens first.

At-Home Activities

How Many Ways?

Gather a collection of small objects, such as paper clips, crayons, pennies, or socks. Start the activity by making a group of 10 objects. Invite your student to make more groups of 10 objects until fewer than 10 objects remain. Together, count by tens (10, 20, 30, 40, 50) and then by ones (51, 52, 53, 54) to find the total. Then, ask your student, "Is there another way to count these objects?" For example, your student may choose to count by ones first (1, 2, 3, 4), and then by tens (14, 24, 34, 44, 54).

Ready, Set, Make 100

Invite your student to play a game with you. Tell your student that each finger represents 10. Begin by making a fist and saying, "Ready, set, make 100!" When you say, "Make 100!" show 0 to 10 fingers. Have your student count how many you are showing. For example, you might show 2 fingers, which represents 2 tens or 20. Then ask your student to show the number of fingers that when combined with yours makes 100. For example, your student would hold out 8 fingers to represent 8 tens or 80 when you are showing 2 fingers. Continue playing and take turns going first.

Adapt this game by using other counting tools to represent 10, such as playing cards or building blocks.

EUREKA MATH²   1 ▸ M6 ▸ TF ▸ Lesson 26

26

Name

1. Draw tens and ones to add.

10 + 75 = **85**	| ||||||| •••••
20 + 65 = **85**	|| |||||| •••••
30 + 55 = **85**	||| ||||| •••••

For 10 + 75, I draw 1 ten. Then I draw 7 tens 5 ones.

There are 8 tens 5 ones, so 10 + 75 = 85.

I can move 1 ten from 75 to show 20 + 65. I still have 8 tens 5 ones, or 85.

For 30 + 55, I can move another 10. I still have 8 tens 5 ones, or 85.

All the totals have 8 tens 5 ones because we just moved a ten to the other part, or addend.

2. Make **95** with two cards.

Show how you know.

| 70 | 15 | 80 | 90 | 5 | 45 | 50 | 25 |

$\underline{90} + \underline{5} = \underline{95}$ $\underline{80} + \underline{15} = \underline{95}$

$\underline{70} + \underline{25} = \underline{95}$ $\underline{50} + \underline{45} = \underline{95}$

I look for pairs of numbers that add to make 95.

I know that 90 + 5 = 95. I draw a number bond to show this.

I use the arrow way. 80 and 10 more is 90. 90 and 5 more is 95. $80 \xrightarrow{+10} 90 \xrightarrow{+5} 95$

I use a number bond to break apart 25 into 20 and 5. 70 and 20 make 90. 90 + 5 = 95.

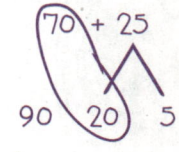

The only cards left are 50 and 45. I draw 5 tens to show 50. I draw 4 tens 5 ones to show 45. 9 tens 5 ones is 95.

Name

1. Draw tens and ones to add.

10 + 35 =

20 + 25 =

30 + 15 =

2. Make **85** with two cards.

Show how you know.

| 70 | 25 | 50 | 80 | 5 | 15 | 60 | 35 |

_____ + _____ = 85   _____ + _____ = 85

_____ + _____ = 85   _____ + _____ = 85

EUREKA MATH²   1 ▸ M6 ▸ TF ▸ Lesson 27

27

Name

1. Add.

   Show how you know.

   24 + 36 = **60**

   20 4   30 6
   20 + 30 = 50
   4 + 6 = 10
   50 + 10 = 60

   > I break apart both numbers into tens and ones to make adding easier. I add the tens. I add the ones. I add 50 and 10.

   > I draw both numbers as tens and ones. I compose a ten with 10 ones. Now there are 9 tens 3 ones. So, 58 + 35 = 93.

2. Add.

   Show how you know.

   58 + 35 = **93**

   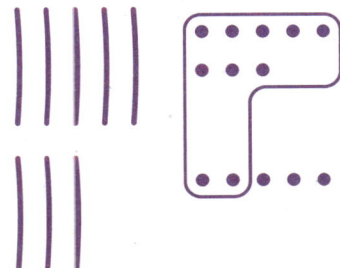

REMEMBER

3. Write the time on the clock.

Circle the time that matches the clocks.

(4 o'clock)   half past 4

> The hour hand is on the 4.
> The minute hand is on the 12.
> I write the time as 4:00.
> It is 4 o'clock.

4. Add.

Show how you know.

46 + 7 = 53

> I can make the next 10 using a number bond.
>
> I know 46 needs 4 more ones to make 50.
>
> I can get 4 from 7.
>
> I break apart 7 into 4 and 3. I make 50 by adding 46 and 4.
>
>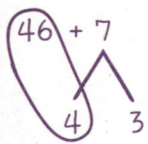
>
> I add 50 and 3 to find the total.
> 50 + 3 = 53

5. How many more apples are there than oranges?

There are 2 more apples than oranges.

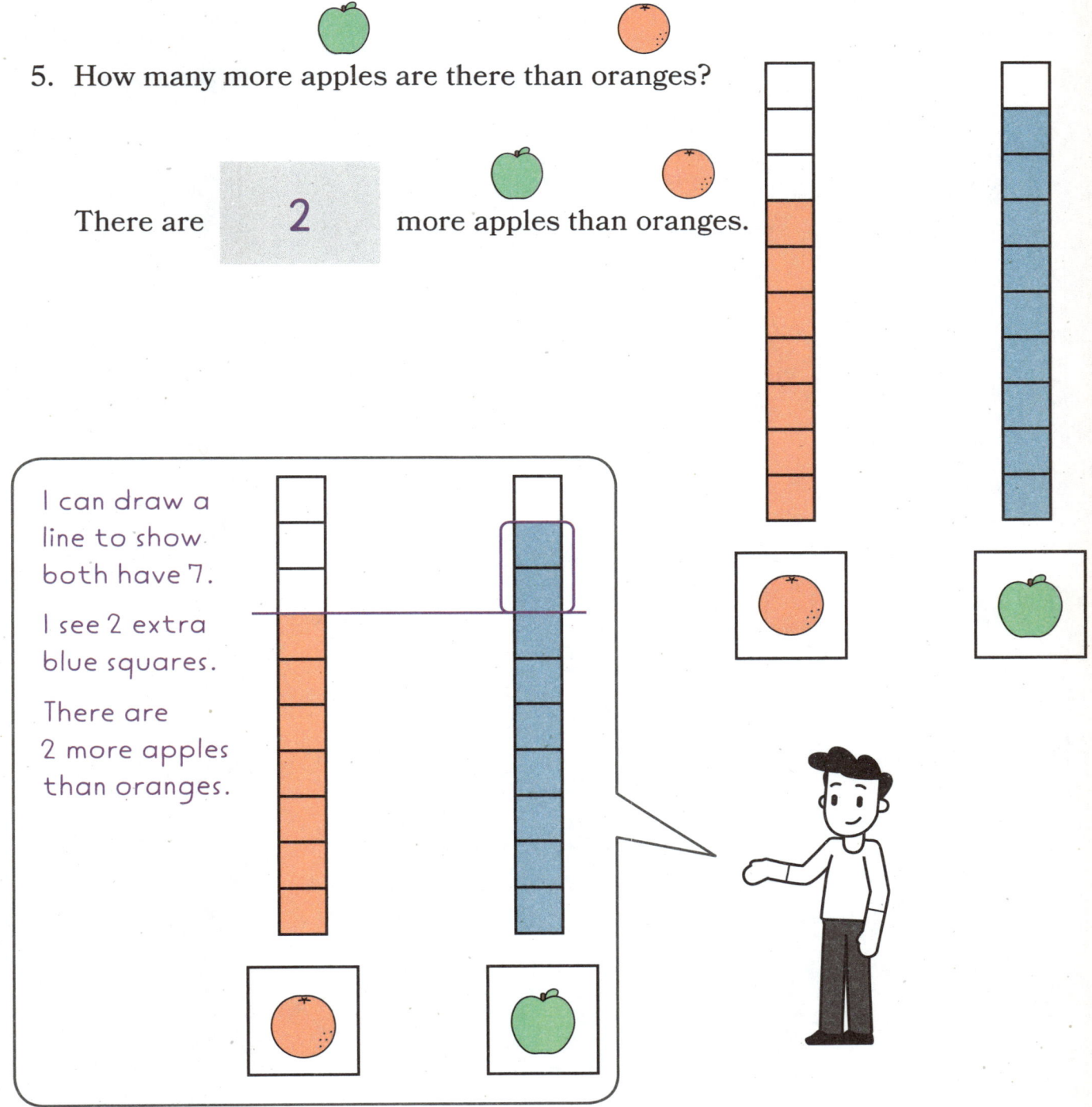

I can draw a line to show both have 7.

I see 2 extra blue squares.

There are 2 more apples than oranges.

1. Add.

   Show how you know.

   12 + 18 =

   55 + 34 =

   44 + 37 =

REMEMBER

2. Write the time on the clock.

 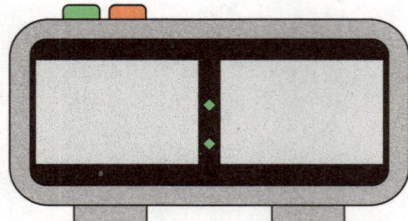

Circle the time that matches the clocks.

6 o'clock   half past 10

3. Add.

Show how you know.

27 + 8 =

4. How many more dogs are there than cats?

   There are _____ more dogs than cats.

EUREKA MATH²   1 ▸ M6 ▸ TF ▸ Lesson 28

Name

1. Add two ways.

$$52 + 24 = \boxed{76}$$

70 + 6 = 76

52 + 24 = 76
       /\
    20 4
52 + 20 = 72
72 + 4 = 76

I can draw tens and ones.
52 has 5 tens 2 ones.
24 has 2 tens 4 ones.
There are 7 tens and 6 ones. 70 and 6 is 76.

I can break apart one number and add the tens first.
I break apart 24 into 20 and 4.
I add 52 and 20 to get 72.
I add 72 and 4 to get 76.

129

2. Add two ways.

$$32 + 58 = \boxed{90}$$

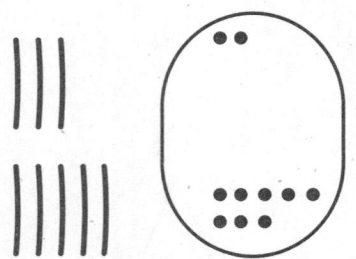

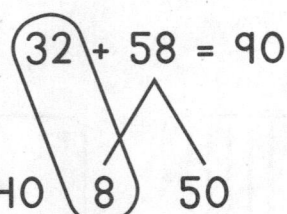

I can draw tens and ones.
32 has 3 tens 2 ones.
58 has 5 tens 8 ones.
I circle 10 ones to compose a ten.
I see 9 tens, or 90.

I can break apart one number and make the next ten to add.
32 needs 8 more to make 40.
I break apart 58 into 8 and 50.
I add 8 and 32 to get 40.
I add 40 and 50 to get 90.

Name

Add two ways.

34 + 45 =

23 + 37 =

58 + 27 =

1. **Read**

   Peg makes 5 jars of pickles.

   There are 10 pickles in each jar.

   She makes 5 more jars of pickles.

   How many pickles does Peg have now?

   Draw Sample:

   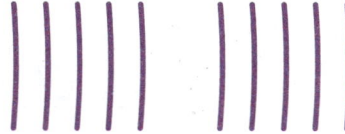

   I read the problem. I read again. As I reread, I think about what I can draw.

   I draw 5 tens to show 5 jars of 10 pickles.

   I draw 5 more tens to show 5 more jars of 10 pickles.

   I need to figure out how many pickles Peg has now.

   I add tens to find the total: 5 tens + 5 tens = 10 tens. 10 tens is 100.

   Write

   50 + 50 = 100

   Peg has 100 pickles.

2. **Read**

Baz has 10 boxes of muffins.

There are 10 muffins in each box.

Baz sells 7 boxes of muffins.

How many muffins does Baz still have?

Draw Sample:

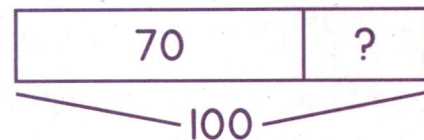

I read the problem. I read again. As I reread, I think about what I can draw.

I draw a tape diagram.

10 boxes of 10 is 100, so I write 100 for the total.

7 boxes of 10 is 70, so I write 70 in one part.

I need to figure out how many muffins Baz still has.

I can subtract tens to find the unknown part: 10 tens − 7 tens = 3 tens.

3 tens is 30.

Write

$$100 - 70 = 30$$

Baz has 30 muffins.

REMEMBER

3. Write the total.

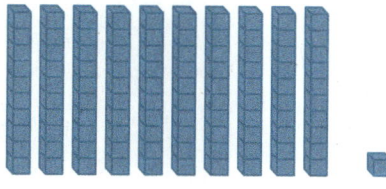

Total **101**

I count 10 tens.
I know 10 tens is 100.
There is 1 one.
10 tens 1 one is 101.

4. Circle how we measure.

We measure the length with cubes.
We make sure there are no gaps.
We also line up the cubes with both endpoints.
We find the length by counting the cubes from endpoint to endpoint.

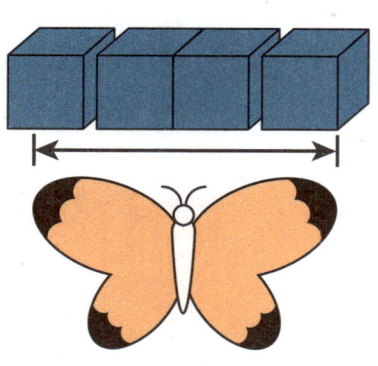 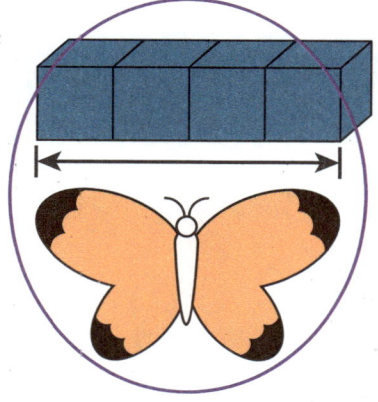 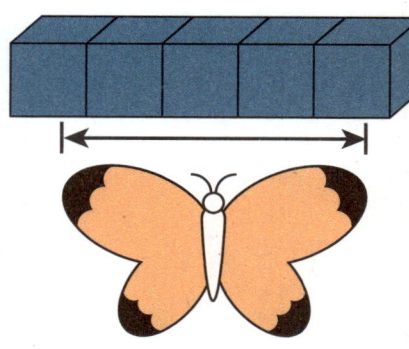

Write the length.

4 centimeters

EUREKA MATH²

Name _____

1. **Read**

   Max has 2 boxes of pencils.

   There are 10 pencils in each box.

   He buys 8 more boxes of pencils.

   How many pencils does Max have now?

   Draw

   Write

   Max has _____  pencils.

2. **Read**

   Beth has 10 boxes of oranges.

   There are 10 oranges in each box.

   Beth sells 4 boxes of oranges.

   How many oranges does Beth still have?

   Draw

   Write

   Beth has _____  oranges.

REMEMBER

3. Write the total.

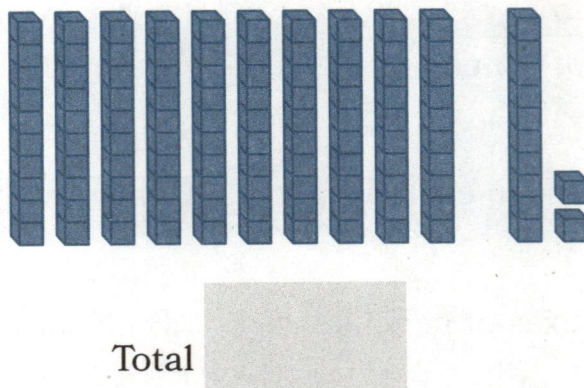

Total

4. Circle how we measure.

Write the length.

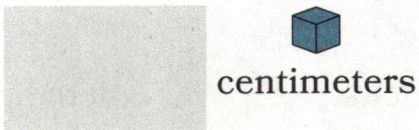 centimeters

EUREKA MATH²   1 ▸ M6 ▸ TF ▸ Lesson 30

Name

1. Make 100 with the number path.

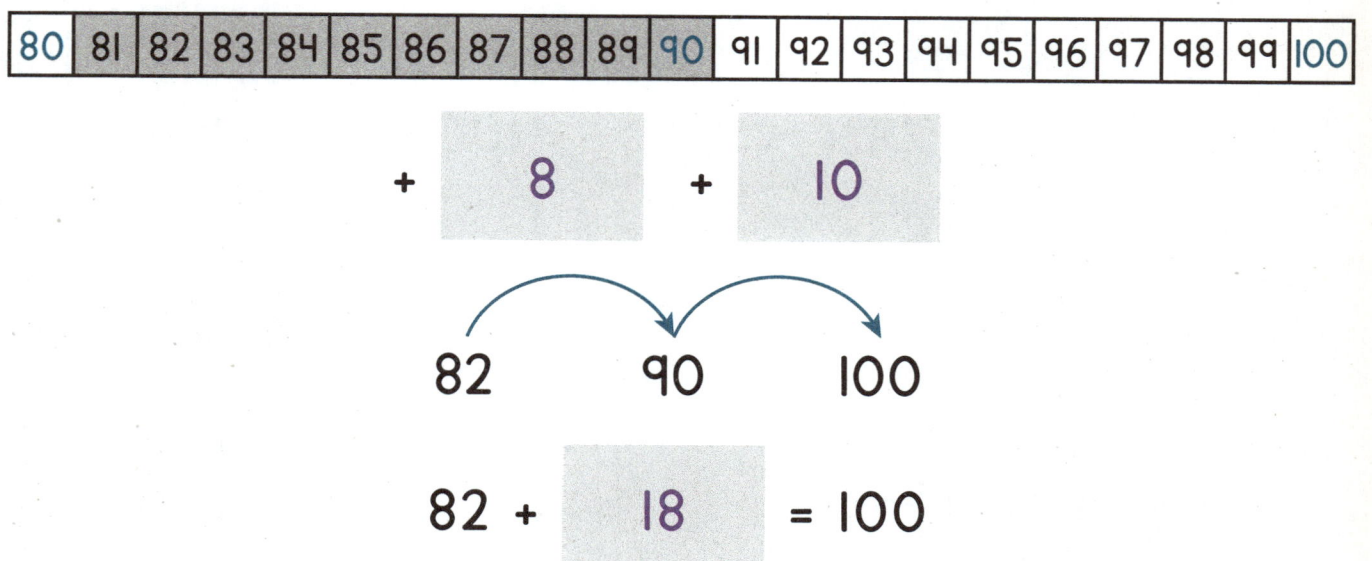

I start at 82.

I can make the next ten. I hop 8 to 90.

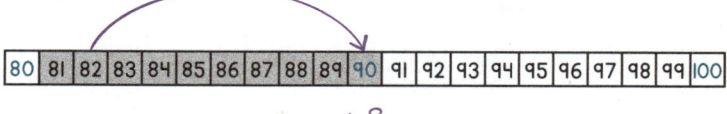

I hop 10 to 100.

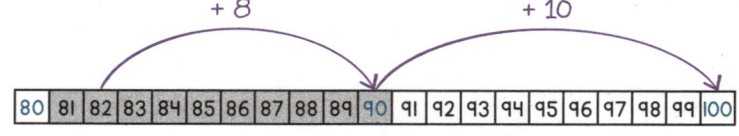

I add the hops: 8 + 10 = 18.

82 + 18 = 100

2. Find the unknown part.

    Show how you know.

   79 + 21 = 100

   I start at 79.
   I know that 1 more is 80.
   $$79 \xrightarrow{+1} 80$$
   I know that 2 more tens is 100.
   $$79 \xrightarrow{+1} 80 \xrightarrow{+20} 100$$
   I counted on 21.
   $$79 + 21 = 100$$

EUREKA MATH² 　　　　　　　　　　1 ▸ M6 ▸ TF ▸ Lesson 30

Name _____

1. Make 100 with the number path.

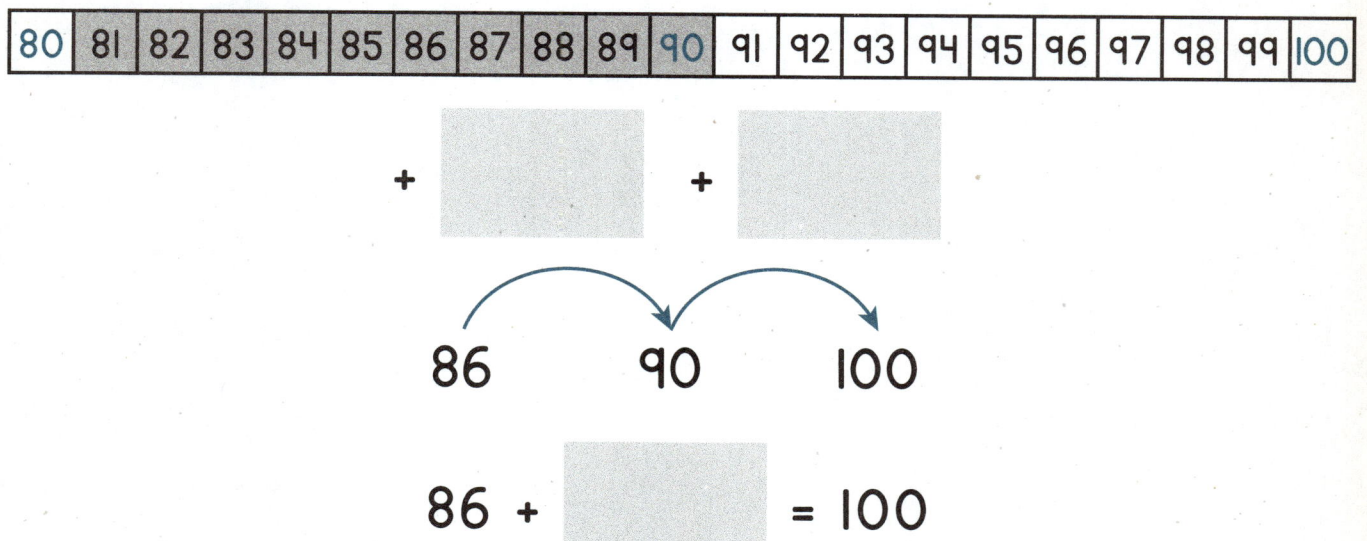

2. Find the unknown part.

   Show how you know.

   62 + ☐ = 100 　　　|　　　 57 + ☐ = 100

EUREKA MATH²   1 ▸ M6 ▸ TF ▸ Lesson 31

Name

1. Add.

   Show how you know.

   $60 + 40 = \boxed{100}$

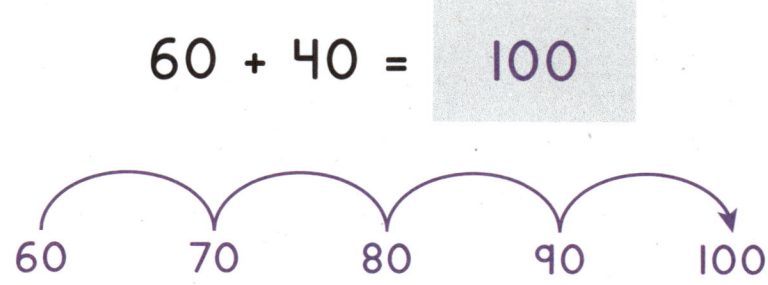

I can count on 40 from 60.
I count on 4 tens.
$60 + 40 = 100$

REMEMBER

2. Count up.

   Write the numbers.

   103, 104, 105, __106__, __107__, __108__, __109__, __110__

   37, 47, 57, __67__, __77__, __87__, __97__, __107__

Patterns show me how to count.
I see the numbers go up in order.
I count by ones to finish the first pattern.

+1 +1 +1 +1 +1
105, 106, 107, 108, 109, 110

I count by tens to finish the second pattern.

+10 +10 +10 +10 +10
57, 67, 77, 87, 97, 107

EUREKA MATH² 　　　　　　　　　1 ▸ M6 ▸ TF ▸ Lesson 31

Name

1. Add.

   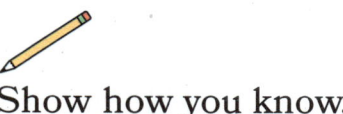

   Show how you know.

   10 + 90 = ☐ 　　　　　 ☐ = 40 + 50

   5 + 95 = ☐ 　　　　　 ☐ = 51 + 51

REMEMBER

2. Count up.

   Write the numbers.

   113, 114, 115, _____, _____, _____, _____, _____

   42, 52, 62, _____, _____, _____, _____, _____

Acknowledgments

Kelly Alsup, Lauren Brown, Dawn Burns, Jasmine Calin, Mary Christensen-Cooper, Cheri DeBusk, Stephanie DeGiulio, Jill Diniz, Brittany duPont, Melissa Elias, Lacy Endo-Peery, Scott Farrar, Krysta Gibbs, Melanie Gutierrez, Eddie Hampton, Tiffany Hill, Robert Hollister, Christine Hopkinson, Rachel Hylton, Travis Jones, Kelly Kagamas Tomkies, Liz Krisher, Ben McCarty, Maureen McNamara Jones, Cristina Metcalf, Ashley Meyer, Melissa Mink, Richard Monke, Bruce Myers, Marya Myers, Andrea Neophytou Hart, Kelley Padilla, Kim L. Pettig, Marlene Pineda, Elizabeth Re, John Reynolds, Meri Robie-Craven, Robyn Sorenson, Marianne Strayton, James Tanton, Julia Tessler, Philippa Walker, Lisa Watts Lawton, MaryJo Wieland

Trevor Barnes, Brianna Bemel, Adam Cardais, Christina Cooper, Natasha Curtis, Jessica Dahl, Brandon Dawley, Delsena Draper, Sandy Engelman, Tamara Estrada, Soudea Forbes, Jen Forbus, Reba Frederics, Liz Gabbard, Diana Ghazzawi, Lisa Giddens-White, Laurie Gonsoulin, Nathan Hall, Cassie Hart, Marcela Hernandez, Rachel Hirsh, Abbi Hoerst, Libby Howard, Amy Kanjuka, Ashley Kelley, Lisa King, Sarah Kopec, Drew Krepp, Crystal Love, Maya Márquez, Siena Mazero, Cindy Medici, Ivonne Mercado, Sandra Mercado, Brian Methe, Patricia Mickelberry, Mary-Lise Nazaire, Corinne Newbegin, Max Oosterbaan, Tamara Otto, Christine Palmtag, Andy Peterson, Lizette Porras, Karen Rollhauser, Neela Roy, Gina Schenck, Amy Schoon, Aaron Shields, Leigh Sterten, Mary Sudul, Lisa Sweeney, Samuel Weyand, Dave White, Charmaine Whitman, Nicole Williams, Glenda Wisenburn-Burke, Howard Yaffe

Credits

For a complete list of credits, visit http://eurmath.link/media-credits